The Times of Their Lives

Women, Men, and the Clock and Watch Industry in Bristol, CT, 1900-1970

Philip Samponaro

Library of Congress Cataloging-in-Publication Data

Samponaro, Philip.
　The times of their lives : women, men, and the clock and watch industry in Bristol, CT, 1900-1970 / Philip Samponaro.
　　　pages cm
　Includes bibliographical references and index.
　ISBN 978-0-9823584-8-1
　1. Clock and watch makers--Connecticut--Bristol--History--20th century. 2. Clock and watch making--Connecticut--Bristol--History--20th century. 3. Clocks and watches--Connecticut--Bristol--History--20th century. I. National Association of Watch and Clock Collectors. II. Title.
　HD8039.W22U673 2013
　681.1'13097462--dc23

　　　　　　　　　　　　　　　　　　2013041470

The Times of Their Lives: Women, Men, and the Clock and
Watch Industry in Bristol, CT, 1900-1970
© 2014 by the National Association of Watch & Clock Collectors, Inc.
ISBN No. 978-0-9823584-8-1

All rights reserved. No part of this publication may be stored in a
retrieval system, reproduced, or transmitted in any form by any means,
electronic, mechanical, photocopying, recording, or otherwise,
without written permission from the publisher.

Printed in the United States of America
The National Association of Watch and Clock Collectors, Inc.
Editor: Diana M. DeLucca
Associate Editors: Freda Conner and Amy L. Klinedinst

Requests to use material from this work should be directed to:
The National Association of Watch and Clock Collectors, Inc.
514 Poplar Street, Columbia, PA 17512

Founded in 1943, the National Association of Watch and Clock Collectors, Inc. (NAWCC)
is a nonprofit member organization whose purpose is to encourage and stimulate interest
in the art and science of horology for the benefit of NAWCC members and the public.

See the last page of this book for more information about the NAWCC
and a reproducible membership application.

All images in this publication are courtesy of the author unless otherwise noted.
The content in this book has been previously published as a series of articles in the
NAWCC *Watch & Clock Bulletin,* Issue Nos. 376, 396, 397, 398, 399.

Dedication

For Bristol's people,
past, present, and future,
and especially
Dorothy F. Beaucar

About the Author

Philip Samponaro is associate professor of twentieth-century United States history at the University of Texas at Brownsville. His interest in clockmaking stems from his family, who for four-and-a-half generations engaged in timepiece manufacture in Bristol, CT.

About the Cover

The painting by Dorothy F. Beaucar featured on the cover is titled *The E. Ingraham Company, Bristol, Ct - Year 1919*. Born in 1919, "Dot" is a lifelong resident of Bristol, and she worked at Ingraham and its successors for 50 years prior to her retirement in 1987. She painted this work as a gift to the author in August 2006. Dot is shown here with the painting in June 2007.

Acknowledgments

This book owes its existence to the contributions of many. The project started as a dissertation in the History Department at the University of Connecticut at Storrs. There, former Connecticut State Historian Christopher Collier provided valuable insights into the local contexts of Bristol's history through his expertise on the state's past. Bruce M. Stave helped me to understand the value and possibilities of oral history. My advisor, the late Susan Porter Benson, provided careful guidance and overall brilliance in developing the dissertation in exciting and new ways. Sue's example as a teacher, scholar, and human continues to guide me today. Betsy Pittman and the staff of UConn's Thomas J. Dodd Research Center greatly facilitated access to relevant collections, which made my employment there as a student archivist especially meaningful during the days of dissertation research and writing. As fellow grad students, friends, and scholars, Charles McGraw, Margaret Robinson, and Rosa E. Carrasquillo offered helpful critiques through their attentive readings of drafts. In the subsequent transformation from dissertation to book, the manuscript and its arguments also benefited from several anonymous readers' reports from historians at different institutions. More recently, the project profited from colleagues in the History Department at the University of Texas at Brownsville. David Fisher, together with his family, provided much encouragement as did Amanda Taylor-Montoya, whose critique of an earlier version of the introduction helped me to frame it in its current state.

The idea to publish the book with the NAWCC came through a happy and chance email exchange in 2011 with John Reardon, the noted Patek Phillippe expert and a native of Bristol, CT. John's suggestion held more merit than he may have known. In the fall of 2007 I had the privilege of addressing the NAWCC at that year's Ward Francillon Time Symposium. I spoke on the subject of the twentieth-century workplace in Bristol's clock and watch industry, a presentation that member Ken DeLucca played a major part in arranging. I then worked with Ken's wife, NAWCC editor Diana DeLucca, in publishing the talk as an article in the October 2008 issue of the *NAWCC Bulletin*. Diana was such a professional and first-rate editor in that project that I embraced the opportunity of working with her in bringing this book to publication. Diana has not let me down; she and her staff have skillfully handled the preparation of the manuscript and accompanying photos better than I could have hoped for, and a special word of gratitude to Freda Conner, who did the indexing over the course of an entire summer. Thank you, NAWCC publications staff, for your careful attention to, and sincere interest in, this book.

Bristol-area friends and contacts also deserve gratitude. Bob Robles of Bristol arranged an interview with his grandfather whose memories of Ingraham's case shop provided the basis for understanding the CIO's campaign at the factory. Son of a beloved production superintendent at Sessions, John Mastrianni Jr., of Forestville arranged most of my interviews from that factory. Similarly, Dorothy "Dot" Beaucar of Bristol provided the majority of names on the interviewee list for Ingraham. Through the years, Dot has been the undisputed nexus of Ingraham retirees in Bristol and at age 94 in 2013 remains a dear friend with whom I am often in contact. I am honored that she painted the cover work for this book.

Finally, I am grateful to my family. My parents taught me and my siblings about the rich history of Bristol, my mother's hometown, when we were still young and nurtured a respect and love for the city in our frequent trips there from our nearby home. My wife Felicia and son Roy have been a steady source of support. Both have shared me with this project for too long, unselfishly giving up time that should have been spent with them, to allow the completion of the work. Felicia and Roy, this book belongs to you as much as anyone.

Table of Contents

Introduction 1

Part I: Class, Family, and Work 5

Chapter 1: An Industry Culture of Work:
  The Local Business through Mid-Century 7

Chapter 2: The Men and Women of the Factories 27

Part II: Change and Its Meaning 41

Chapter 3: Workers Take a Stand: CIO Unions at Ingraham and Sessions 43

Chapter 4: Tempus Fugit: Livelihoods in Trouble 57

Epilogue: Time Runs Out: The End of Work 71

Notes 73

Bibliography 83

Introduction

"By its clocks, Bristol regulates the world." So went a popular twentieth-century saying to describe Bristol, CT. Since the nineteenth century, the city had been the focal point of the national clock and watch industry, which from its inception had claimed Connecticut as its center. So renowned was the local timepiece industry that Bristol, about 20 miles southwest of Hartford, earned the title of "The Clock City." Outsourcing and urban renewal programs during the 1960s brought an ignominious end to that distinction and the bustling factory life that had once characterized the city.[I.1]

Figure I.1. Bristol, CT, Main Street, late 1930s.

As a youth growing up nearby, I tried to rekindle the images of workers and bosses pouring into the streets after another workday as I traveled through the increasingly empty streets of post-industrial 1980s Bristol. What happened to those men and women, whose lives gave meaning to the local industry? This is their story.

What follows is a study of the social relations of labor in the American clock and watch industry during the twentieth century, particularly changing family and workplace cultures at two companies in Bristol: the E. Ingraham Company and the Sessions Clock Company. The latter dominated the Forestville section of Bristol. Both were family-owned businesses and together formed the backbone of Bristol's timepiece industry during the twentieth century. The story begins at the start of the century when Sessions, formed in 1903, joined Ingraham in shaping Bristol's clockmaking sector. It concludes seventy years later for two reasons: (1) By 1970 outside interests replaced the local families in running each firm; (2) 1970 is the year the clock and watch industry reached its nadir in both Bristol and the United States, having fallen victim to both internal and external pressures—with Sessions closing altogether. The industry became a memory, leaving workers to find new jobs in an increasingly service-oriented economy.

I argue that the clock and watch industry did not benefit from participation in the military-industrial complex. Unlike many other examples in labor and business studies of the twentieth century, clock and watch workers identified exceptionally closely with their workplaces prior to the 1950s. Workers were content with employers' empowering programs of steady work and high wages and accepted the CIO unionism of the early 1940s not so much as labor militancy but as a continuation of the "family" ideology that prevailed both inside and outside the factory walls. Government involvement with the industry, especially at the federal level, undermined the foundations of worker identification, whittling away at job security on a long-term and permanent basis. Transformation from civilian to defense production, a process that occurred during the Cold War, made the local industry, particularly Ingraham, dependent upon the production of antiaircraft fuses for survival. At the local level, bosses failed to find alternatives that could counter the detrimental effects of federal policies, further weakening the economic position of workers from the 1950s onward.

The organization of work and, within it, workers' identities and how identities evolved over time, frame the story. Ingraham and Sessions employed men and consistently high numbers of women after 1900, which is unusual when compared to other factories, where women's workplace numbers did not increase until World War II. However, within the steadily maintained gender division of labor, worker identity changed, together with the bases of division and privilege in the workforce. Early in the century men claimed breadwinner wages as theirs alone, but Cold War ideology, with its emphasis on middle-class conformity and family consumption, extended the ideal of a family wage to women. Engineers, once rare in the industry, became increasingly visible as designers of military ordnance and saw themselves as respectable professionals, even while remaining firmly in working-class jobs. By the end of the 1960s, blurring working-class categories was commonplace among other workers, including women. Class segmentation was also affected by the changing relationship of different ethnicities and races.

Figure I.2. The E. Ingraham Company plant on North Main St., Bristol, ca. 1935.

The workforce included Italians, Irish, French Canadians, Germans, and others. At the very end of the period, the introduction of nonwhites under equal opportunity programs imparted a more pronounced sense of who was and was not a legitimate worker, based on race. Labor activism at the end of the period confirmed the persistence of a strong working-class identity that survived the industry's demise.

Gender was a consistent factor in evolving worker identity. Workers at Ingraham and Sessions forged bonds within industry, family, community, and even economy. These crucial links allowed them to define their jobs and environments separately from those of their bosses, and more fundamentally, to survive both from day to day and in times of economic crises, such as depressions and the rare strikes that occurred. Called work cultures, these sources of power are not fixed but vary even within the same factory according to type of employment, sex, age, ethnic and religious affiliation, and larger cultural traditions. As with race, ethnicity, and nationality, there is, in fact, a multiplicity of gendered identities, defined through an evolving process in people's lived experiences. One retiree, Lillian Rock, described herself as a "girl," a rank and file worker without authority, when talking of her work as a married employee at Ingraham during the 1950s. At the same time, she also recognized that at home she was both a mother and head of her household, explaining how she exercised authority over her small children while her husband worked the nightshift elsewhere. Lillian held at least two different gender identities, between which she negotiated daily.[1,2]

Forged under the paternalist, or family-oriented, management style of owners Ingraham and Sessions, the factories outwardly followed a patriarchal family model, in which managers and bosses were to be heeded like fathers by their employees. Women were expected to act like good daughters or "girls," to draw upon Lillian's words, working docilely and dutifully before leaving jobs when they got married, to then become society's wives and mothers (although, as Lillian's example suggests, few women could afford, or even wanted, to follow this route). Working men were to behave as sons, or apprentices of sorts, respecting bosses often many years their senior and learning from them the path to the breadwinner status to which they aspired.

On the surface, the workplace appeared to have been a static and even restrictive arena, representative of how company or industry cultures are often conceived. Like examples elsewhere, paternalism affected relations between not only men and women but also between skilled workers and apprentices, between older and younger men, and even between ethnic groups. Women suffered from the patriarchal structure of the workplace, but so too did younger and more junior men, and age differences and divisions shaped attitudes toward unions. Older workers with long service records, for example, often rejected unionism in favor of the possibility of continued paternalism, doing so out of a sense of family-like loyalty to their workplace and regardless of gender. In contrast, younger employees with shorter records, especially those men who felt pressed upon by male bosses above them, rallied behind the union. It is indeed remarkable how deeply imbedded the patriachalism and family metaphors cultivated by the owners were within work culture and industry. However, changes constantly caused the renegotiation of genders within the workforce. The transitions to unions and later corporate ownership opened new paths to workers, altering patriarchal norms and allowing men and women to challenge established patterns and to carve out new spaces in the workforce in line with the times.

Public and scholarly interest in timepieces has all too often emphasized the timepieces themselves with little or no concern about attention to the hearts, hands, and heads behind them. On occasion, enthusiasts and professionals have looked at the early entrepreneur/mechanics

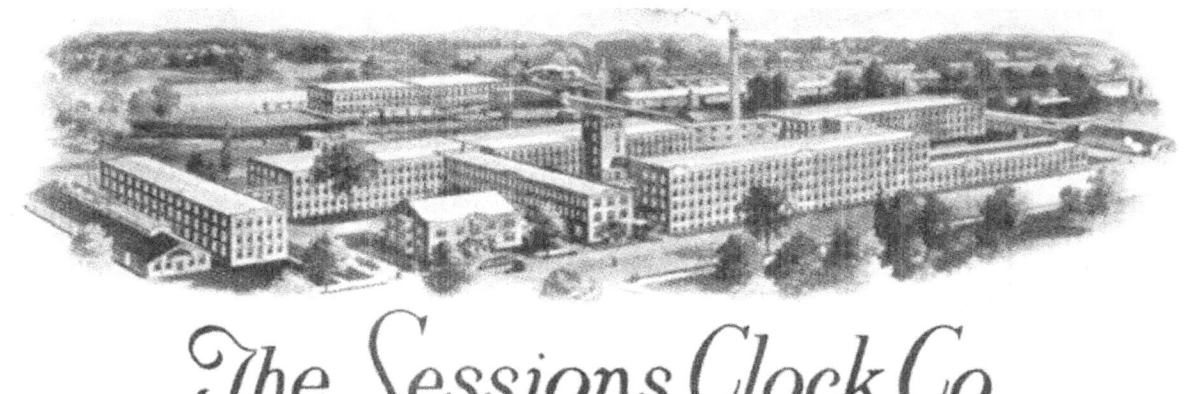

Figure I.3. The Sessions Clock Company, ca. 1950.

who revolutionized manufacturing techniques during the nineteenth century but hardly gave a thought to factory hands. Thus, other than looking at the bonds that developed between clockmakers and apprentices who went on to make names for themselves in the trade, no real understanding of the social relations of production has been achieved for the clock and watch industry.[I.3]

By telling this story, I hope to fill in gaps in relevant literature on the clock and watch industry. Before undertaking this project, I was puzzled why a formal study of the social relations in the industry had not already existed. It seemed odd that no labor historian had focused on the men and women who produced the timepieces that proliferated during this period. By measuring time, clocks and watches serve to organize the processes of production in a capitalist society, making the absence of such a work curious indeed.[I.4] Three decades ago, Roy Rosenzweig showed that, in spaces where time is technologically conditioned, workers have sought to separate time for leisure from that of work in order to have some control over their lives. In Rosenzweig's seminal account, wage earners in Worcester, MA, between 1870 and 1920 separated their time on the job from their time off in the effort to compensate for their subordination to machines on the shop floor. Recognizing the meaning of their own time, these laborers zealously guarded their autonomy in "leisure time," fending off various efforts by local industrialists to extend control from the factory into this "arena," which in turn conditioned their own workplace battles with the employer and his clock. For Rosenzweig's workers, time consciousness was a base of power.[I.5]

The same year that Rosenzweig's work appeared, economic historian David Landes published a memorable study on the role of clocks in shaping the modern world. Landes aimed to define the relationships between the precision of time patterns and moving parts by studying the history of time from the Middle Ages forward. In a provocative text, he followed the model of emphasizing timepieces and the innovators behind them without studying the development of the division of labor or time management in general. Landes did take his discussion into the twentieth century and talked about labor in the watchmaking industry of England and United States. His focus here, however, was not changing work hours or the rise of "work" and "leisure" time but rather technological advances in watchmaking.[I.6]

So the door remained open to write about workers and their times in the clock and watch sector. I was excited by the prospects of getting to meet the veterans of this workforce and using their voices to tell the story of such an acknowledged center of clockmaking in the United States as Bristol. So many men and women who worked in the factories still lived in the 1990s that the possibility of writing the social history of so important an industry was easily within grasp. Chris H. Bailey, horologist and longtime curator at the American Clock and Watch Museum in Bristol, unknowingly encouraged me with a comment he wrote in an editorial for the Winter 1996 issue of the museum's quarterly, *The Timepiece Journal*. "All too often," Bailey reasoned, "the 20th century is ignored by students of horological history, yet a great deal of interesting data is available and even some of the 'old timers' are still around."[I.7] I began to envision a study that did for Bristol what the groundbreaking *Brass Valley: The Story of Working People's Lives and Struggles in an American Industrial Region* (Philadelphia: Temple University Press, 1982) had accomplished for the nearby Naugatuck Valley. In that social history, authors Jeremy Brecher, Jerry Lombardi, and Jan Stackhouse showcased the words and reflections of the working men and women of Waterbury and Ansonia, CT, to recapture the industrial world of Naugatuck Valley for the century after 1880. I hoped to create descriptions of real people who came alive through their shop-floor/kitchen table perspective, much as the authors of *Brass Valley* had done.[I.8]

My first priority was to interview as many women and

men who made the timepieces as possible. I made contacts within the retiree community in Bristol and then conducted oral histories with them and their friends, a process that occurred in the fall of 1996 and spring of 1997. When finished, I had interviewed 13 men and 10 women from Ingraham, whose service records collectively spanned the years 1918 through well after the end of the period, and 5 men and 13 women from Sessions who worked at various times from the later 1930s onward. I had representative interviews of skilled and unskilled workers, management and labor, and many different ethnic groups, including Yankees (or Anglo Americans), Irish Americans, Italian Americans, Polish Americans, and German Americans who made up Bristol's workforce prior to 1970. Unfortunately, my sampling was not so successful on the subjects of unions, which played a major role in the local industry after 1940. There, despite making every effort to get a representative sample, I found only a few former workers who had been either union members or even pro-union. Union records helped to fill in the gaps of organized labor's story. So, too, did the company histories and industry records with which I paired workers' accounts of their experiences.

I realized that life at Ingraham and Sessions was made up of several interrelated stories. One is a community or industry account that focuses on sex-segregation, gender-based work cultures, and gender identities of workers, often complicated by differing ethnicities. Another is the story of the clockmaking industry in Bristol as affected by local players and government tariff policies. A third is the account of labor generally, and especially of the unions that workers fought to win and maintain. All are essential to understanding what was once Bristol's premier industry.

As a business and labor history of the timepiece industry, this study compares two different firms and management strategies within the framework of the transition from paternalist and welfare capitalism to labor unions and ultimately outside corporate ownership, together with the effects on employer/employee relationships. I explore the role that family ownership played in shaping gender and policy at the two companies, dissecting its effects on both the development of managerial strategies and the companies' arguments for government tariff protection of the timepiece industry nationwide. The inclusion of later strategies from outside managers provides a useful comparison of very different styles of running the business and the worker responses to them, particularly in response to the transformation from civilian to defense production that occurred during the Cold War.

This study also addresses the everyday lives of working men and women involved in the making of timepieces, a topic otherwise ignored in the study of the U.S. clock and watch industry for any period of its history. The findings challenge dominant historical models about men and women's relationships to waged labor by revealing new variations of those ties within several distinct frameworks. These include, in the case of Ingraham, one of the more empowering forms of welfare capitalism yet studied in the United States; unstudied variants of CIO unionism; a new vision of wartime production; and, ultimately, the permanent reconfiguration of the industry to defense manufacturing. Most of all, the story of Bristol's clockmakers reminds us of the need to resist fitting different groups of industrial workers within a single dominant historical paradigm.

In the effort to adequately assess continuity and change in Bristol's timepiece industry, this study is subdivided into two parts. The first part considers family and work patterns at Ingraham and Sessions. Chapter 1 looks at the industry culture of work, addressing the paternalist policies of the family owner/managers as well as labor trends in working-class families. Gender-based hierarchies that characterized the social structure of individual departments within the factories are discussed in Chapter 2. Part 2 examines the evolution within the cultures of family and work. Chapter 3 tackles the issue of unionization from its initial success through the destructive divisiveness of the early Cold War. Chapter 4 examines worker life amid changes wrought by outside managers in the last decades of the industry history, in the process assessing worker identity at the very end of that history. The epilogue assesses the legacy that workers and their industry left to Bristol.

PART I
Class, Family, and Work

Chapter 1

An Industry Culture of Work: The Local Business through Mid-Century

At the start of the twentieth century, brothers William S. and Walter A. Ingraham stood as the third generation of employers to head the family firm of the E. Ingraham Company since its founding in 1831. Three years later, fellow Bristol industrialist William E. Sessions took over Bristol's other clockmaking concern, the E. N. Welch Manufacturing Company, located in the Forestville section of town, and renamed it the Sessions Clock Company. Over the next several decades, the companies remained family-owned. Likewise, they shared the distinction of being at the heart of the national clock and watch industry, historically centered in Connecticut and with Bristol as its top producer.

Despite these commonalities the fortunes of the two firms differed. Plagued with financial burdens, the smaller Sessions Clock Company followed a turbulent path, sometimes turning a profit with its popular clocks but often in economic trouble and requiring federal assistance for survival during the Depression. In contrast, Ingraham took a more stable route. Unlike Sessions, Ingraham also made inexpensive pocket watches. The widespread appeal of these products in the 1930s allowed the firm to expand its workforce considerably even during the Depression to become one of the largest timepiece factories in the United States. When World War II engulfed the nation, both companies shifted production to war-related products and afterward returned to peacetime production. By the 1950s, the winds of deindustrialization began to sweep over Connecticut's cities, eroding the stability of the local clockmaking sector. In this climate, the owners had second thoughts about the futures of their companies and ultimately allowed the sales of the firms to corporate interests.

The story of Sessions' survival, paired with the history of growth and prosperity at Ingraham, and the ultimate decline of both spanned five decades. During that half century, workers and owners developed a unique culture that was understood differently by both groups. As established civic leaders, benefactors, and leading employers, the Ingrahams and Sessionses promoted the city's welfare by including the workplace in their sense of civic commitment through a program of welfare capitalism. They saw themselves as heads of patriarchal "family" units and made decisions on behalf of workers for their own good, even if contrary to the latter's wishes. The owners passed their businesses from fathers to son, and maintained paternalistic oversight of their workforce across generations. At Ingraham and Sessions, paternalism was outwardly well-intentioned and included steady work and high pay. It was also a policy of social coercion used by the owners

Figure 1.1. Elisha. N. Welch, founder of the E. N. Welch Manufacturing Company, Sessions Clock's predecessor.

Figure 1.2. Elias Ingraham, founder of the E. Ingraham Company.

to quell class conflict and expand the capitalist base of their businesses while enforcing employees' paths for development and growth in a very hierarchical atmosphere. Unlike other cases of welfare capitalism, the balance between coercion and benevolence in Bristol tipped in favor of the latter, empowering workers in the decades before unionism by the Congress of Industrial Organizations (CIO) in the late 1930s.[1.1]

Beneath the owners, workers had their own notions of what family and community meant and pursued their own goals at the factories. Employees amassed long service records and brought in other workers who were either related or neighbors. Family and kin mediated the experience of work and community for a workforce that was multiethnic and often segmented along ethnic lines. The multifaceted sense of community bridged ethnic diversity in the workplace and extended back into heterogeneous neighborhoods. In the process it forged a counterpoint to the family image cultivated by the owners and ultimately abetted the unionization of both factories in 1941.

Origins and Trajectory of the Clock and Watch Industry in Bristol

The companies around which workers and their bosses built their world were manufacturers of the product that had brought fame to Bristol. The use of interchangeable parts in the manufacture of wooden movements by entrepreneur Eli Terry in the early nineteenth century in nearby Plymouth, local clockmaker Joseph Ives' invention of roller pinions and the introduction of rolled brass in movement making in the 1830s, and then Chauncey

Jerome's local manufacturing of a cheap 30-hour brass movement beginning in the later 1830s revolutionized the domestic clock industry and put Bristol in the forefront of U.S. clock production by midcentury. Aided by a steady source of power in the Pequabuck River, which ran through the centers of both Bristol and Forestville, the entrepreneurial successes achieved in the city encouraged many commercial ventures into clockmaking. In the century after the start of local clockmaking in the 1790s, an estimated 200 individuals, partnerships, and concerns manufactured clocks in Bristol with varying degrees of success.[1.2]

Despite the proliferation of clockmakers in Bristol, many of the firms were small-scale operations and competed vigorously for a scarce local supply of essential materials such as brass and clock springs, items that would not be available in bulk until after the founding of the Bristol Brass Company in 1850 and spring maker Wallace Barnes Company more than a decade later. The tensions between small size and fierce competition reduced the number of manufacturers. With fewer employers, the industry became increasingly concentrated under the factory system after 1850. By 1880 only three major clockmaking concerns were left, all of which operated as factories: Ingraham (then called the E. Ingraham & Company), the E. N. Welch Manufacturing Company, which made low-priced clocks and movements, and its offshoot, Welch, Spring, and Company, which produced higher-grade clocks such as calendar and astronomical timepieces. With the absorption of Welch, Spring, and Company by E. N. Welch in 1884, only Ingraham and Welch remained. Together, Ingraham and Welch—after 1903 Sessions Clock Company—constituted the local timepiece industry during the twentieth century.[1.3]

For the most part, the men and women who worked at Ingraham and Sessions came from Bristol and surrounding areas. If not from Bristol proper, workers most often lived in the two neighboring communities of either Terryville, which shares the city's western border and is within a few miles of the downtown Ingraham factory, or Plainville, running along Bristol's southern border where Forestville is located. Plainville's immediate proximity to Forestville ensured Sessions Clock a steady stream of workers, especially after the inclusion of the town on Bristol's trolley route with the establishment of the Bristol and Plainville Tramway Company in 1893.[1.4] Terryville's relative proximity to Ingraham brought locals to that factory regularly from the 1930s onward when busses replaced trolleys and traveled to Terryville (as well as Plainville). Trains also brought workers as far as 20 miles away, including from Waterbury, and after the post-1945 automobile revolution from such distant towns as Hartford, Torrington, Winsted, Canton, and Avon.

At Ingraham, company founder Elias Ingraham (1805-1885) provided popular clock designs such as the sharp Gothic, Venetian, and Doric models that allowed his firm

VENETIAN
One-day or eight-day pendulum springwound time and strike
1860-70

Figure 1.3, above. Sharp Gothic clock designed by Elias Ingraham, ca. 1845-1850.

Figure 1.4, right. Venetian and Doric clocks designed by Elias Ingraham.

DORIC
One-day or eight-day pendulum springwound time and strike
1870-80

to prosper. As president from 1885 to his death in 1892, Elias' son Edward (1830-1892) pushed production ahead with the invention of the black enameled wood case. Thereafter, the company became a premier producer of clocks cased in a style that became the standard finish for high-grade mantel clocks in the U.S. clock industry until about 1930. Such success greatly affected employment numbers at the factory. In 1880 when Bristol had a population of 5,347, the company employed 125 hands, the fourth largest employer in Bristol. By 1920 Ingraham's employment number stood at 884, or more than seven times its 1880 level. Proportionately, Ingraham grew faster than Bristol, whose population of 20,620 reflected only a four times increase from 1880. Employment numbers soared during the decade to 1,284 by 1928, making it second in size only to the New Departure Division of General Motors, the giant ball bearing maker, which operated in the city with some 3,112 employees by April 1928. Ingraham's employees grew in number from 1,368 in 1930 to 1,893 in 1935 and then to 2,387 by 1940, doubling during the decade.[1.5]

Before he died, Elias left explicit instructions that his offspring expand his company under the Ingraham name. In a letter to a grandson on Christmas Day 1884, the deeply religious elder charged Edward's son Irving "with this Express wish and Expectation that you are to Settle down in lifetime Business and Ucefilness in Church Society and Town." "And Now," the patriarch commanded, "I Want

Figure 1.5. Main Office, E. Ingraham Company, February 16, 1914. William Shurtliff Ingraham seated at right, third row.

Figure 1.6. William Ingraham (center), with brothers Walter A. (left) and Irving (right), Bristol, ca. 1895.

this Clock Business to be kept and continued in the Hands and Home of the Ingraham Family for Many Generations to Come."[1.6] As employment numbers show, successive generations of Ingraham males heeded his order. The firm advanced its product line several times, first under another of Edward's sons, William S. (1857-1930), who as general manager and treasurer oversaw daily operations of the company, and then under William's sons Edward (1887-1972) and Dudley (1890-1982), who became president and vice-president, respectively, upon their father's retirement in 1927—and, in true paternalist fashion, at his direction. In 1897 the firm added cheap one-day alarm clocks to its traditional production of 30-hour and 8-day movements and in 1915 instituted the manufacture of 8-day alarms. Looking to expand into the market for cheap, nonjeweled watches, the company engineered the buyout of the Bannatyne Watch Company in Waterbury, at a receiver's sale in 1912 and brought its equipment to Bristol. A year later, it began the manufacture of cheap pocket watches, whose low-grade (and cost) earned the popular product the name "dollar watch." An electric automobile clock entered the Ingraham catalog in 1929 and inexpensive men's wristwatches in 1930. In 1931 the firm started the manufacture of self-starting electric clocks, followed by manual-start

Figure 1.7. "The E. Ingraham Co. Clock Shop by Night, Bristol, Conn.," 1920s.

Figure 1.8, left. An Ingraham watch ad, ca. 1930s.
Figure 1.9, above. Self-Starting Synchronous Motor Alarm Clocks, Ingraham catalog for 1934-1935.

electrics and radio cabinets in 1932, and electric motors and timers beginning in 1934. To round out its line, in 1938 the firm added women's nonjeweled (those lacking the lubricated jewels that prevented friction and wear between metal pinions and plates) wristwatches. Because of the Depression, the expansion into watches was especially profitable, enjoying a high demand during the 1930s because of their low price; "clock watches" accounted for 38 percent of Ingraham's production totals by 1934, and prof-

its from them helped the firm to expand the physical size of its plant between 1933 and 1937 by 75 percent. Moreover, they were largely responsible for Ingraham's claim to 27.5 percent of the entire domestic output of all clocks and watches by 1940.[1.7]

Sessions Clock was less lucky, treading in the footsteps of its predecessor E. N. Welch. Despite success under founder E. N. Welch, the earlier company declined rapidly after his 1887 death and went into receivership in October 1893. Although it reopened in 1897, the firm suffered two separate fires in 1899 that destroyed both the movement and case shops and by 1902 was on the verge of bankruptcy. The accumulation of stock first by Bristol businessman John Humphrey Sessions (1828-1899), interested in establishing his family in the clock industry, and then his offspring, allowed the family to control the firm by January 1903. Sessions' son William Edwin (1857-1920) assumed the presidency before his son William Kenneth Sessions, Sr. (1887-1969) took over after his father's sudden death in 1920. Under "W. E.," Sessions Clock operated as a profitable enterprise, making 2,000 clocks a day by 1907 and trebling its production capacity in the 17 years of his tenure.[1.8]

After W. E.'s death, the firm enjoyed modest product success but was also mired in financial problems.

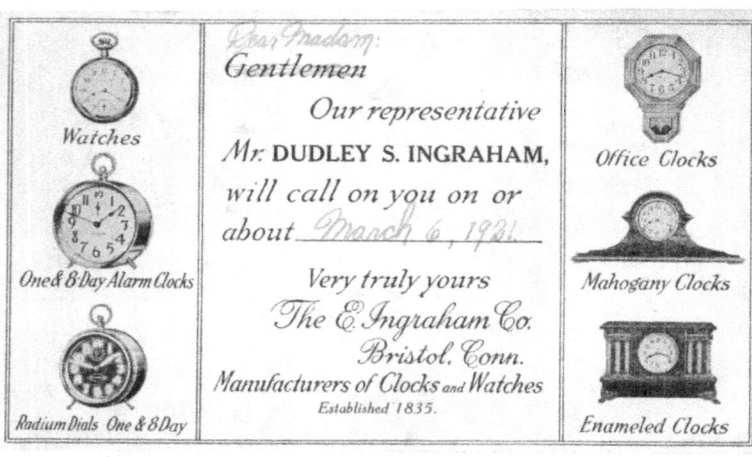

Figure 1.10. A postcard from Dudley S. Ingraham, 1921, showing clock styles typical of both Ingraham and Sessions in the early 1920s.

Figure 1.11, left. William Edwin Sessions. *Bristol Press*, January 5, 1905. **Figure 1.12, right.** William Kenneth Sessions, Sr. From *Sessions Clock Company, 127 Years of Clock Craftsmanship: Clock Timers, Clock Movements, Timing Controls*, inside front cover. AMERICAN CLOCK AND WATCH MUSEUM, BRISTOL, CT.

William Kenneth, or "W. K." introduced a popular duet clock that struck two notes for each hour during the early 1920s, and in 1927 a successful two-train Westminster chime, which was designed as a less expensive alternative to the traditional three-train movements of these models that sold as luxury items. Despite the start of the Depression, the firm adopted the synchronous electric clock system in 1930 and made an electric strike and Westminster chime within a year. Constant experiments in motor and gear design led to the adoption of a hysteresis motor, invented by Bristol entrepreneur, Arthur Haydon, which when mated to a worm gear-driven movement produced one of the quietest movements of the day. So popular were Sessions' electrics for their quietness that they accounted for 80 percent of production totals by 1941, up from 57 percent in 1931 and a mere 13 percent a year previous.[1.9]

The firm, however, shouldered heavy unpaid city taxes as the outcome of a combination of unpaid inheritance taxes at W. E.'s sudden death in 1920 and the negative effects of the 1921 depression. The inability to pay these taxes all at once led to a mounting load of more unpaid taxes that increased with reduced overall production after 1929 and even further after a disastrous first attempt to manufacture an electric motor in 1931 and 1932. Collectively, these circumstances led to losses between 1931 and 1933 and ultimately to $45,000 in unpaid city taxes by 1934. Sessions managed to survive only by securing a loan from the Federal Reconstruction Finance Corporation (RFC) for $140,000. To help to secure stability, in 1938 the company added a popular line of mechanical timers for use in cooking, photography, or any application requiring an audible signal at the end of a preset time. Likewise, from 1940 to 1947 the firm derived income by renting factory space left empty by Depression-wrought cutbacks to the Haydon Manufacturing Company, manufacturers of timing devices and motors for Sessions' electric clocks.

Because of these measures, Sessions Clock weathered destabilizing effects of the Depression on the national timepiece industry, particularly before the mid-1930s. Between 1899 and 1929, the industry remained fairly constant in size, with 79 firms at the beginning and 81 at the end, despite a 10-year high of 91 firms between 1909 and 1919.[1.10] With the onset of the Depression, the number of firms dropped, together with employees and wages paid.[1.11] In Connecticut, bigger clockmaking concerns than Sessions, such as Gilbert and Ingersoll-Waterbury, went into receivership. In response to such cases, the National Recovery Administration enacted a Code of Fair Competition to ensure continued industry success. Aimed at controlling work hours and production totals, the code was grudgingly signed by industry leaders—including Republican Edward Ingraham, who sat on the negotiating committee—on February 26, 1935.[1.12] Stability ensued and continued after the NRA was declared unconstitutional, despite some loss due to the recession of 1937 (Table 1.1). While prevailing in this environment, Sessions remained small as a consequence of its troubled past. In contrast to the mammoth Ingraham a few miles away, the struggling Sessions employed 415 in 1940, or less than a pre-Depression figure of 487 in April 1928.[1.13] These figures corresponded to the low percentages of production totals at Sessions in comparison to those at Ingraham and the larger United States clock and watch industry through 1942, when civilian manufacture ceased (Table 1.2).

During World War II, both Ingraham and Sessions, like many other industrial firms, performed their patriotic duty as manufacturers of government ordnance. Sessions made its primary contribution by supplying wooden chests, cases, boxes, and cabinets to several large defense industries. The company also made a variety of nonwood products, including bullet cores, disks for motor oil filters, and exhaust stacks for planes in lightweight duraluminum. Similarly, Ingraham manufactured numerous small metal items for the military, such as millions of bullet cores, firing pins, and filter plates, but those products were few in comparison to its major contribution of antiaircraft fuses. Prior to U.S. entry into the war, Ingraham's pocket watch program had provided the firm the opportunity to experiment with these small clocklike devices that were in short supply in the early 1940s. Ingraham

Table 1.1 Vital Statistics of the U. S. Clock and Watch Industry, 1929-1939						
Total	1929	1931	1933	1935	1937	1939
Firms	81	75	53	77	75	74
Employees	21,450	16,213 (-24.4%)	12,850 (+20.7%)	18,047 (+40.4%)	23,223 (+22.3%)	17,878 (-23%)
Wages in Dollars	25,750,08 8	15,685,61 5 (-39.1%)	10,003, 392 (-36.2%)	18,307,99 2 (+83.0)	27,558,69 9 (33.6%)	20,467,333 (-25.7%)

Source: Department of Commerce, Bureau of the Census, Washington, "Census of Manufactures: 1935: Clocks and Watches," January 25, 1937, 1, Ingraham Company Papers; and "Clocks, Watches, and Materials and Parts (Except Watchcases)," *Sixteenth Census of the United States: 1940: Manufactures: 1939, Vol. II, Part 2* (Washington, D.C. 1942): 317-319, 317.

Table 1.2
Production Totals of the U.S. Clock and Watch Industry, Sessions Clock Company, and the E. Ingraham Company

Year	Pieces, Total Industry	% of Total Industry, Sessions Clock	% of Total Industry, Pieces, Ingraham
1930	15,659,917	2.0	22.8
1931	13,345,647	1.4	27.1
1932	11,480,242	2.2	30.2
1933	15,000,000*	2.0	31.2
1934	18,400,000*	1.3	31.4
1935	21,260,113	2.0	29.8
1936	27,380,411	1.6	29.5
1937	26,239,025	1.7	28.5
1938	20,495,736	1.3	28.3
1939	25,185,809	2.2	27.1
1940	28,863,811	2.5	27.5
1941	33,288,680	856,005	26.2
1942**	11,288,680	301,241	25.5

Source: Chris H. Bailey, "Sessions Clock Company's Annual Production, 1930-1953," in the reprinted *Sessions Clocks Catalogue No. 65, 1915* (Bristol: American Clock and Watch Museum, 1977), 117; Ingraham Company, "All Available Statistics on Clock and Watch Industry," author's collection.
*Estimate.
**First two quarters only.

created the first mass-producible antiaircraft fuse in history and thus emerged as a key contractor for the War Production Board. This pivotal position won the firm two Army-Navy "E" Awards for excellence in war production in June and November of 1944.[1.14]

World War II was a mixed bag for the local clock and watch industry. On one hand, it breathed new life into Sessions and guaranteed continued work for Ingraham. But the war did not bring higher wages and increased employment that happened elsewhere, nor did it enhance the industrial success of the companies, in comparison to other industries for which the government built and fitted out plants that were later converted to civilian production. Rather than following the model of expansion and postwar subsidy associated with other industries, these companies, especially Ingraham, operated with a drastically reduced workforce under tight government controls, which undermined their position in the local economy. Despite full conversion to the war effort, much of Ingraham's factory space and machinery went unused: for example, 118 of 403 automatic screw machines, critical for making the components that held fuses together, lay idle in July 1944.[1.15] Unable to realize full capacity, Ingraham could not take on additional government contracts. Moreover, as a result of federal bureaucracy, a protracted conversion to civilian production after the war prevented a quick return to prewar production levels.

Federal tariff and tax legislation from the prewar era did even less to serve the clockmakers' interests. The Federal Excise Tax, enacted in 1941 and continued after the war, instituted a tax of 10 percent on key industries such as automobiles, clocks, and watches to generate funds for government preparedness programs. Meanwhile, the government had instituted a tariff policy in the mid-1930s that favored imports from European nations faced with the threat of Nazi aggression, namely, Switzerland, which had long been a center of timepiece production, particularly watches.[1.16]

The clock and watch industry had enjoyed historically high tariffs. A 1922 tariff, for example, safeguarded the industry from a flood of German imports priced 25-50 percent below American merchandise of the same quality. With the onset of the Great Depression, Congress passed an emergency tariff in 1930 that instituted high rates on all imports. Most manufacturers were pleased, especially Ingraham, which thrived in the closed market of the 1930s. Ingraham's fortunes, and those of the entire industry, changed in 1935 with the passage of the Hull Reciprocal Trade Agreements, named in honor of their sponsor, Secretary of State Cordell Hull. These acts allowed the president to make a treaty with any country, and the rates agreed upon with that particular country automatically applied to every other country, with the exception of Germany, which had been debarred by presidential proclamation. A Hull agreement with Switzerland went into effect on February 15, 1936. Through a combination of low wages both in factories and the cottage industry, Switzerland hit the American market hard; within a year the imbalance of all trade between the United States and Switzerland shifted from two to one in 1934 to three to one in favor of Switzerland. In 1937 watches constituted 40.5 percent of all timepiece imports, slamming Ingraham especially. At that time, the percentage of Ingraham's payroll on wristwatches had declined to 13.2 percent from 19.4 percent two years earlier, corresponding with a decline during the same period in the percentage of the national industry making these items from 20.5 percent to 16.1 percent.[1.17]

Figure 1.13. Antiaircraft fuse, manufactured by Ingraham, early 1950s.

As part of an effort to create a bulwark against fascism in Europe, Congress upheld protection on Swiss imports as war loomed. For the domestic industry, this decision had swift consequences when U.S. entry into the conflict cut off watch production at home. With the watchmaking sector completely converted to manufacture after June 1942, Switzerland filled the large demand for watches by civilians and servicemen. Watch imports increased regularly, giving Switzerland a monopoly on the U.S.

market through 1945 and allowing it to establish such a dominant base in sales and brand identity it proved a challenge for American firms to compete. Even when imports ultimately proved insufficient to meet demand and the wartime Office of Price Administration (OPA) set maximum prices for watches, the Swiss overcame this obstacle with innovative technology and increased production to sustain profit. Manufacturers' slow postwar conversion paralleled the flood of more inexpensive Swiss timepieces into the market. Ingraham, still reeling from wartime controls, suffered doubly in this process. By 1946 Ingraham produced only 13.9 percent of the industry's total watches and clocks, down from more than 25 percent prior to the war, and earned only 8.35 percent of industry dollars versus roughly 17 percent in the prewar economy. Sessions fared proportionately poorly, making only 212,340 timepieces in 1946 out of the national total of 27,991,531.[1.18]

Aggravating the situation, by 1950 consumers wanted wristwatches, more easily accessible and considered more "modern" in an increasingly high-paced world, than the cumbersome, old-style pocket watch for which Ingraham was known. Consumer desire for watches in general skyrocketed far more than the demand for clocks after 1945, with market consumption growing 320 percent from 1935 to 1950. This figure had a bleak underside for Ingraham because it did not include the nonjeweled pocket watches that at one time had formed the backbone of the company's financial success. Washington remained unmoved by such a circumstance due to the high domestic demand. Consequently, by 1950 the Swiss had cornered 65 percent of the American watch market—in contrast to 23 percent only 15 years earlier (Table 1.3). To make matters worse, the post-war Congress also facilitated trade with Japan, where the United States hoped to rebuild a democratic partner in capitalism. This move proved further disadvantageous domestically because of low Japanese wages that cut product costs significantly. Ingraham, for example, averaged $1.87 per hour in wages in 1948 against a shocking 9 cents per hour paid by the largest producer of clocks and watches in Japan, a differential that made competition impossible.[1.19]

The excise tax and tariff policy produced a strong need for lobbying and public posturing by the paternalist Ingraham and Sessions families, especially after the former was enlisted to help support the Cold War defense buildup; standing by 1954 at 20 percent on all timepieces

Table 1.3 Percentages of Watch Consumption by American Market, 1935-1951			
	1935	1940	1950
Swiss	23%	44%	65%
American Jeweled	20%	19%	14%
American Non-Jeweled	57%	37%	21%

Source: Dudley S. Ingraham to Messrs. Edward M. Greene, Edward T. Carmody, November 21, 1951, Edward Ingraham Papers, I, 9.

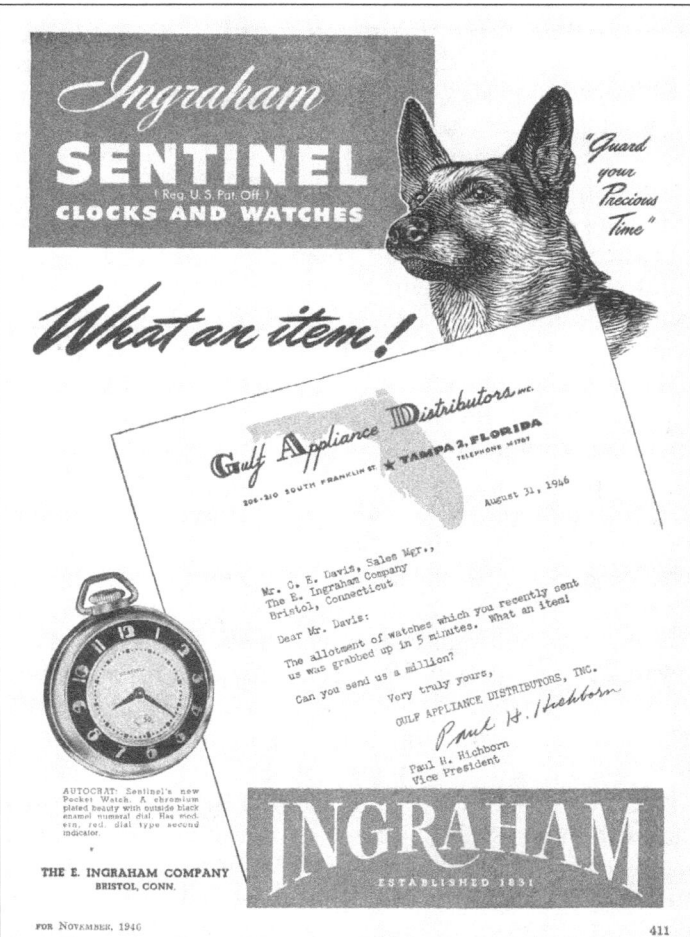

Figure 1.14. Ingraham Sentinel ad, 1946.

except for alarm clocks that sold for under $5 (the latter were taxed at 10 percent). An outspoken opponent of both the tax and tariff, Dudley Ingraham spearheaded the fight, and by the end of the 1940s quickly won recognition as the industry spokesman, especially on tariffs, in part through his membership in the American Tariff League, an organization dedicated to "a sound and adequate tariff policy for the protection of American industry, agriculture and labor." Such efforts helped to effect changes in 1954, with the national cessation of a new round of defense contracts to supply antiaircraft fuses for the Korean War. First, Congress reduced the excise tax to a uniform but still damaging 10 percent on clocks and watches under $65. Concurrently, in June 1954 President Dwight Eisenhower introduced a higher tariff, which increased rates as high as 50 percent on the critical market of imported watches with no jewels, or up to 17 jewels. Restoring these tariff rates on watches to pre-1935 levels, Eisenhower's action reflected both recommendations from the Tariff Commission and also the administration's belief that watch companies were essential to national defense. The changes, however, were too little too late; by April 1956 Connecticut had lost 3,700 of the 11,700 jobs that it had in the clock and watch industry in 1948 and Bristol 416 of 2,816.[1.20]

The Owners, Civic Leadership, and Paternalism

Despite such hardships, the owners acted as father figures to their workers, especially prior to unionization. For example, during the Depression W. K. Sessions allowed 10 of the 23 apartments that Sessions Clock owned to go rent-free and leased the remainder at reduced rent. In the more prosperous 1920s, he reminded his employees of his outward good will toward them by continuing the annual September barbecue begun by his father. Throughout his administration, W. K. remained in close everyday touch with his workers, touring the plant regularly, and even attending the Christmas parties the workers held at the factory. All the while, of course, his agenda included an underlying premise of keeping his workforce loyal.[1.21]

In his civic gestures toward his employees, W. K. emulated his father, who had styled himself as a paragon of community activism. Like other paternalist employers, W. E. Sessions played a major role in public life. He spearheaded the establishment of banks and trusts, businesses, and civic organizations, and was a trustee at Wesleyan University. He also played a hand in local politics; a lifelong Republican, Sessions was a burgess, along with William Ingraham, when Bristol had a borough district from 1893 to 1911, and presidential elector in the national election of 1908. W. E. considered himself foremost a steward of Jesus Christ. Entering the Methodist Church at age 12, he devoted a large portion of his time and money to the institution his father—himself an ardent Methodist—had helped to build up in Bristol. In his lifetime, William donated liberally to the local Methodist Church and served as president of its Board of Trustees, president of the official board, and superintendent of the Sunday school. Fond of children, he applied his religious inclinations outside of Bristol, including New Britain, where he helped to subsidize a children's home. Wanting to spread the ideal of noble living to the young, in 1890 he established and completely subsidized the Mount Hope Methodist Church in the sparsely populated Bristol neighborhood of Chippins Hill, which had no church. For 30 years prior to his death, he served as both Sunday school superintendent and de facto pastor of this community, where he taught and even preached each week after attending his own local parish. He also promoted the promulgation of other Christian denominations in Bristol. Following the incorporation of Bristol's Zion Evangelical Lutheran congregation in 1902, Sessions provided a building lot on Judd Street and even offered plans for the future church.[1.22]

Sessions poured his civic posturing into his capitalist endeavors at the new Sessions Clock Company, believing he had a moral obligation to prevent the closing of Forestville's largest employer and, since the early 1870s, the source of a quarter of its tax base. In March 1904 his company refurbished a dilapidated boarding house into inexpensive and clean apartments, complete with bathrooms, to accommodate deserving workers and their families. Also in 1904 Sessions organized the Forestville branch of the Bristol Library to encourage reading among the local populace and curb less moral pursuits like drinking. The family provided a company-owned building for the library until 1947 and, until the Depression, paid the salary of the librarian. With the completion of a new schoolhouse in 1910, the firm supplied all necessary clocks, which not only helped teach children the meaning and value of time but also reminded them from their earliest schooling of the role played by Sessions Clock in their community.[1.23]

The Ingrahams shared many traits with W. E. Sessions. Likewise, they cultivated an image of a strong commitment to religious activism. A devoted and strict Congregationalist, Elias Ingraham set the example both at work and at home. In 1848 the company founder required that the superintendent of his movement shop sign a contract that included a provision that the latter maintain a morally disciplined workforce and that "[he and his hands] shall be regular attendants at church on the Sabbath." In private life, Elias claimed to uphold a firm belief in a Christian God and lead a clean lifestyle at home, attributes to which he believed he owed his success in the clockmaking business. "It is now," he wrote in his 1884 letter to his grandson, "fifty Seven Years this Evening Since I United my Destiny with Julia H. Sparks." Elias found that he had to apply himself hard to his trade in his early years of marriage to support "One Self and Wife." Such perseverance paid off, thanks, believed Elias, to his faith in God. "But the God," he continued, "in Whom I trust has Blest us in our Union And in our Business and we have lived happily to geather thes[e] fifty Seven Years and have Attained to our present [life]."[1.24]

Elias' offspring followed suit, as his great-grandchildren symbolized. As youth, Edward and Dudley were forcefully conveyed the family sense of civic responsibility emanating from their forbear's examples, both inside and outside the home. Like other members of the Ingraham and Sessions families, they attended public school for part of their education and developed bonds with Bristolites of a very different class, those who would one day work in their factories. "Bertha D.," an anonymous interviewee who inspected 8-day alarms at Ingraham with her mother during the 1930s, recalled the class dynamics of such public school ties, which are often

Figure 1.15. Dudley S. (top left) and Edward (bottom right) Ingraham, with their mother Grace and sister Faith, ca. 1894.

ignored in the historical record. Whenever Edward visited their department, Bertha remembered, he would hug her mother, with whom he had gone to school, and ask how his old classmate was.[1.25]

Such identification no doubt helped to condition the brothers' relationship with and participation in the community. Both men married schoolteachers outside of their social class, and Dudley's second wife, the daughter of German immigrants, was a librarian at the Bristol Public Library. Edward was chairman of that library, vice-chairman of the Bristol Board of Finance, and founder and first president of the Bristol Clock Museum, later the American Clock and Watch Museum. Dudley served as a deacon and president of Sunday school at the local Congregational Church and was also a onetime chairman of the Community Chest, chairman of the local Salvation Army Board, a leading figure in the local Boy Scouts, and a member of the Bristol Board of Education. Both Ingrahams also were active in civic affairs beyond Bristol. Dudley served a term on the Connecticut Highway Safety Board and Edward on the State Planning Board. Edward was also an active supporter of the Connecticut Junior Republic, a center for vocational and citizenship training for boys in Litchfield, CT.[1.26]

As businessmen, Edward and Dudley Ingraham, together with their peer W. K. Sessions, promoted images as fatherly leaders by encouraging the economic growth of their city, region, and industry. All were active on various local boards of directors outside of their clock companies and participated in business activities that were outside of Bristol but tied to the interests of both the city and its clockmaking. All three men were members of the Manufacturers Association of Hartford County and its parent organization, the Manufacturers Association of Connecticut, both of which existed to advance local and state industry and, of course, promote business over labor interests. Likewise, the trio played major roles in the Clock Manufacturers Association of America, a trade organization founded in 1926 to promote the interests of the industry nationwide, culminating in 1949 with the election of Dudley Ingraham as president and Sessions as vice-president of this organization.[1.27]

As did Sessions, E. Ingraham played father to workers, especially during the Depression. To keep short time to a minimum, the firm created new work, especially to protect men's jobs, in line with the dominant notion that males were society's breadwinners. In 1932 Ingraham secured a massively successful contract to build radio cabinets for the infant Emerson Radio and Phonograph Corporation. The purpose was to maintain employment in the mostly male case shop that was undercut by the disappearance of a market for relatively high-priced wood case clocks. Moreover, throughout the Depression the owners approached layoffs in a paternalistic manner by ensuring that if their plant employed more than one member of a family at least one would always be working regardless of any layoffs. Robert "Bob" Tetro, the only non-Ingraham to hold a position in executive management during the family's tenure, confirmed management's commitment to this policy, especially during the 1930s. As Bob recalled, "If layoffs were required, it was very paternalistic. The single men were let go before the married men because the married men had families. The married man had maybe only worked two years and the single man six. That made no difference. Married women were let go [before their husbands]. It was that paternalistic type of thing from the top down." Adhering to paternalism also meant making a tremendous number of small loans to workers throughout the 1930s, maintaining the group life insurance introduced in 1929, and introducing paid vacations and Christmas bonuses for all employees in 1936.[1.28]

Never shy of self-aggrandizement, the company frequently reminded the public, and particularly its workers, of what such treatment meant to a Depression-era workforce and the loyalty from them that it deserved. Typical was a 1940 full-page ad from the *Bristol Press* that Ingraham placed to counter the CIO activism which was then sweeping the city. "Even during the worst part of the depression, 1930-1935," the firm boasted, "the Ingraham Company increased its workforce 38%. Not many industrial plants in the United States can show a similar record. . . . What this has meant to Bristol during the depression years can hardly be realized." The ad attributed this success to the Ingrahams: "Most remarkable of all, since the Company was founded by Elias Ingraham in 1831, 109 years ago, the Ingraham Company has been controlled and managed by his descendants. . . . Unique in American industrial experience is this Ingraham history of close family control coupled with continual growth and expansion.[1.29]

Such haughtily advertised success allowed Ingraham to continue its long-term welfare capitalism program of regularly advancing wages. The foundations for a selective high-wage structure, to which Ingraham aspired, were inherent in the work specific to the clock and watch industry. A clock or watch movement keeps time through the interrelation of separate gear trains that provide precisely regulated releases of energy. Individual parts in a movement needed to be both mass-produced and precision made. Once assembled, movements had to be timed and adjusted to work properly. Skilled work played a heavy hand in all this, and labor accounted for roughly 55 percent of the total selling price of clocks and watches produced in the United States in 1941. As manufacturers of primarily inexpensive, nonluxury timepieces, Ingraham had to reduce this labor cost to ensure low, competitive costs. During the nineteenth century, the company kept costs down like other manufacturers, in part through the continual division of labor. Nevertheless, the skill involved in clock manufacturing remained high enough to prevent the complete degradation of work common to mass production techniques, and the minute division actually increased the value of women's labor, which was classified as unskilled or semiskilled. Moreover, low overheads, coupled with the company's high sales, acted as a buffer against the whittling down of wages and allowed regular advances. As a result, Ingraham by the late 1920s compared favorably to the industry nationwide, which claimed Connecticut as its center, rivaling other large Connecticut timepiece manufacturers such as the New Haven Clock Company and the Gilbert Clock Company in Winsted (Table 1.4).[1.30]

During the 1930s Ingraham's rates meant economic stability for many working families. By decade's end, the average American family of five (two adults and three children) required an annual income of $2,200 per year, or $44 a week, for a modest standard of living. Like other manufacturers, Ingraham did not meet this standard, but its distribution of annual employee earnings put this ideal in reach for men, traditionally seen as society's main sources of income, in certain classifications—if their wives also worked. For example, by 1939 men on production jobs in the movement shop averaged $29 a week while the weekly rate of all jobs for men in that division was $31. With women's plant-wide average at $23, a husband employed in the movement shop and his working wife could theoretically meet the needs of an average household with ease.[1.31]

So high, in fact, were Ingraham wages during the 1930s that a survey by the National Recovery Administration mid decade revealed the firm to pay the highest rates in the entire domestic industry, including the jeweled watch sector; the company was nicknamed "the Ford of the industry." Locally, the firm's wages interfered with the ability of New Departure to operate, ultimately forcing Bristol's chief employer to adopt a new pay structure to compete with Ingraham. Contributing over $26,000,000 to the Bristol economy between 1928 and 1938, Ingraham emerged as the city's most financially secure industrial employer during the 1930s, a distinction the owners sought to uphold. The Ingrahams, particularly Dudley, who had legal training, campaigned actively against competitive threats to their wage structure, first from cheap domestic prison labor and, after the U.S. reciprocal trade agreement with Switzerland in 1936, from inexpensive imports.[1.32]

The owners had their own vision for the workforce. They wanted to create a homogenous body of competent and stable workers, whose lifestyles benefited both Bristol and their companies—a type of worker whom contemporary theorist Antonio Gramsci described in his discussion of Fordism as "suited to the new type of work and productive process."[1.33] Under that scenario, they would be kind fathers, so long, of course, as the workers behaved.

To bolster this objective, both the Ingrahams and the Sessions stayed in constant contact with their workers, especially prior to unionization. Because of the owner's accessibility, wage earners readily identified with them. Former workers talked fondly of their employers. Sandy Giammateo, who started at Sessions in 1941, recalled how much W. K. Sessions' presence at workers' Christmas parties meant to her. "[Where] would you get a president of a company that would [come to] your Christmas party?," she exclaimed. "Not today you don't see that. They have their own personal private parties." Flattered by Sessions' attention, Sandy and her fellow workers showed him their appreciation. "How Mr. Sessions used to love to sing," she recollected. "We all sang, we made his day." "It was like a big happy family," said Sessions retiree Catherine "Kitty" Baldaccini. "I used to enjoy going to work. I couldn't wait!" The feelings at Ingraham were no less strong, as longtime watchworker Kathryn "Kay" Laviero recalled when thinking of Edward Ingraham, who impressed her with his kindness when she was a young woman in the 1930s:

> I was coming down the stairs and going home and it was raining so hard, and here I'm standing, I'm looking up at it. He [Edward] just came out and he looked at me and he says, "You got an umbrella,?" and I said, "No umbrella." "Do you want a ride home?" I said, "Yes, I'd like a ride home." He gave me a ride home. Yeah, he owned the factory. He gave me a ride home.[1.34]

The close, family-like identification of workers with their employers was particularly strong during the Depression because the firms offered steady work. With the widespread lack of jobs elsewhere, employees of both companies expressed loyalty and gratitude for whatever work they had. Decades later, former workers vividly remembered how lucky they felt to be earning wages vis-à-vis less fortunate others. One of the first memories Bertha D. recounted was of looking down from the factory windows with other female co-workers during the Depression as unemployed men from New Departure walked by with their heads down, "weeping without work." This sight

Table 1.4
Average Hourly Rates, March 1928

Department	Ingraham	Sessions	Gilbert	New Haven
Machine Screw, Men	.662	.6145	.615	.82, .66, .73
Press, Men	.46	.509	.49	.65
Wood Rubbers, Men	.6172	.753	.64	.67
Alarm Clock Assembly, Women	.35	.37	.35	.46
Timing, Men	.75	.7424	.55	.66
Plate Work, Women	.38	.34	.35	.41
Buffing, Men	.796	.574	.65	.666
Wood Assembly,* Men	.605	.614	.66	.63
Machine and Tool, Men	.794	.75	.75	.75

Source: Office memo, March 27, 1928, Ingraham Company Papers, III, 12. Note: wherever the piecework or bonus system were employed, the affected rates were obtained by dividing the amount earned by piece work or bonus by the actual number of hours employed, as secured on a weekly basis.

*This category covered the assembly of movements into wood cases, dialing, and running.

made her realize how fortunate she was to be employed. Such sentiments inspired the roughly 1,500 employees at Ingraham in March 1933 to offer their services without pay if the owners needed them when President Franklin Roosevelt declared a bank holiday, a proposition that the firm's subsequent use of scrip money rendered unnecessary. When not idled by frequent shutdowns, Sessions Clock usually operated on a three-day-a-week schedule except immediately after being granted of the RFC loan, which temporarily returned the workweek to five days. Appreciative for even tenuous work to support their families, employees took on the City of Bristol to keep Sessions from closing permanently. In late September 1934 they demanded city help to facilitate Sessions' receipt of the essential RFC loan by lobbying through their representatives on the Bristol Workers Cooperative, an organization of laborers from local industries set up to promote the welfare of the city's working-class families.[1.35]

Unity with the family owners, both during the Depression and in other periods, demonstrated the success of paternalism. However, while superficially successful, paternalism at both firms failed to overcome the reality of class relations between those who owned and those who worked. Workers broke freely and practically with the limiting policies of paternalism when the latter challenged core values of family and community. Like the shoemakers and tanners of Endicott Johnson, the employees of Ingraham and Sessions achieved what Gerald Zahavi calls a "negotiated loyalty." This loyalty, argues Gerald Zahavi, reflected an understanding of worker rights and obligations in the face of paternalism.[1.36] CIO unions and strikes were the most obvious outcome of this process. The roots of such manifestations were in the day-to-day responses of wage-earning families at Ingraham and Sessions that undermined the owner's model of "family."

Workers As Family Members

The Ingraham and Sessions family model for workers might suggest English-speaking Anglo-Saxon or "Yankees" origins such as the owners. In fact, while management recruited its workforce almost entirely from within Bristol and surrounding towns, a large percentage of their employees were either foreign-born or the children of immigrants. This trend owed its origins to the 30 years after 1890, when Bristol's population, 7,382 that year, nearly tripled, due largely to European immigrants and their offspring. By 1910, when Bristol had 13,502 residents, first- and second-generation Americans accounted for 55 percent of the population, while native-born residents totaled only 36.6 percent.[1.37] As the primary source of workers for Bristol's clock industry, this population translated into an ethnically heterogeneous workforce that mitigated against an easily defined family model. The owners may have hired all-white workforces (workers of color were not regularly employed until the later 1960s), but the workforce was segmented along ethnic lines. Departments were often controlled by members of one ethnic group, and language was divisive on the job and in the community; Kenneth Webster, a former Ingraham watch timer, recounted that when growing up and just starting work at age 16 he lived with his family in an apartment that neighbored one occupied by a French family. Two of his neighbors' daughters also went to Ingraham, but because of language differences kept to themselves. This barrier prevented him and his mother, a watch worker since widowed in 1927, from communicating with their fellow employees.[1.38]

Despite such diversity, by 1920 the work culture fostered employment patterns within families that eschewed ethnic lines, evidenced when comparing the marital status of native-born workers and their immigrant counterparts in the census returns of that year. In line with the owners' emphasis on male household heads, most men, whether married or not, followed the unbroken chain of lifelong employment typical of the owners' equation of breadwinner status with steady work. In contrast, paid labor for native-born and immigrant women happened most frequently among single females between 14 and 29 years old, after which their numbers dropped dramatically, presumably because they left work for marriage or motherhood (Table 1.5).

In 1920 the employment of women in noticeable numbers was a relatively recent but firmly entrenched aspect of factory life. During the nineteenth century, men enjoyed a long-standing dominance in the division of labor in the local clock industry, performing almost all aspects of timepiece production. Women's work, socially acceptable only for the unmarried, was limited to dial and tablet painting. The owners faced a radically different division of work after the introduction, during the late-nineteenth century, of labor-saving foot and hand presses like those used in other "light manufacturing" sectors. Requiring simple, repetitive operation, the presses were seen as suitable for women. Employers coupled this thinking with the growing reliance on elaborate piecework systems,

Table 1.5
Those reporting employment in clock factories, Bristol and Forestville, 1920 by age, sex, and marital status

Age	Men, single	Men, married	Men, widowed	Men, divorced	Women, single	Women, married	Women, widowed	Women, divorced
10-13	1	0	0	0	0	0	0	0
14-19	62	3	0	0	65	4	0	0
20-29	65	43	0	0	68	13	3	1
30-39	21	100	3	1	17	12	1	1
40-49	10	84	2	0	6	8	0	1
50-59	6	68	5	0	2	3	1	0
60-69	0	32	6	2	0	0	0	0
71-79	0	12	6	0	0	0	0	0
80-89	0	0	0	0	0	0	0	0

Source: Manuscript returns for 1920 census, Bristol and Forestville.

which paid by the piece as opposed to an hourly wage. Employing as many women as possible became a key to business success. Jobs were sex-typed; women were given work considered "delicate" by industry standards. These piecework jobs included subassembly, assembly, and finishing clock and watch movements. "Certain operations," concluded Edward Ingraham, a blind believer of this ideology, "cannot well be performed by men, such as the finer assembly jobs, vibrating, finishing, poising, etc." Paternalists who viewed their workers as family members, the Ingrahams and Sessionses applied these strategies and included women prominently in employment figures by 1920. Women remained at least half the workforce thereafter; in 1940 they accounted for 45.9 percent of Ingraham's 2,387 workers. At the smaller Sessions, remembered former production superintendent Francis "Frank" Savage, they represented an estimated 65-70 percent of a workforce of 415. Outside managers maintained these percentages after Sessions was sold in 1957 and Ingraham relinquished family control in 1956, selling in 1967.[1.39]

By paying lower hourly rates to women, Ingraham and Sessions obscured the skilled nature of many female jobs. As producers of complex timing mechanisms with high labor content, both acknowledged the skills of their workforce. "We employ 370 males, 289 females, 636 skilled workers," Ingraham concluded in responding to a financial questionnaire in the early 1920s. In practice, however, management of both factories saw as skilled only those workers engaged in tool and die making, screw production, or adjusting assembled movements. These jobs required relatively few employees, years to master, and were the exclusive purview of men. At Ingraham, this proportion of the workforce approximated the industry average of a mere 13 percent by 1941. Sessions had even fewer "skilled" employees—only 5 percent, according to W. K. Sessions in May 1941. "Almost the entire personnel of the plant," he wrote, "is of the unskilled or only partially skilled such as . . . power and foot press operators, a large proportion of subassemblers and final assemblers, working on a type of product which is not fine precision work." Women performed unskilled labor, or partially skilled jobs at best, but bosses could not deny the talents needed. In its full-page ad in the *Bristol Press* in 1940, Ingraham pointed out the expertise women needed in alarm clock assembly, which required 133 parts and 291 hand operations per clock per assembler. Nevertheless, women were perceived as unskilled because, as supposedly temporary workers, they would not be in the workforce long enough to become "skilled."[1.40]

Such marginalization was not lost upon the workforce at both factories. Co-workers considered women outside the norm if they continued their work for prolonged peri-

Figure 1.16. General Manager W. K. Sessions Jr., and Main Office personnel, Sessions Clock, circa late 1940s. Interviewee Frank Savage is second from left; Dom Dellario (with glasses), another interviewee, is fourth from left in the front row.

ods, especially if married. Women led this practice, often mixing it with an undertone of admiration for a married woman's ability to keep working. At Sessions from 1945 to 1961, Laura (Mastrianni) Santago remembered Ida Moore, a married woman who worked there for almost a half century. Although already retired when Laura started, Moore visited Sessions frequently; she was well-liked but still viewed as odd because of her service record. Laura found herself in a similar situation when her employment continued for some years, even after her own marriage, and faced comparisons to Moore. "People used to say, 'Gee, are you still working in Sessions Clock?,' and I used to say, 'Yeah, I'm still there.'" They then taunted Sally jokingly, "What are you going to be, another Ida Moore?"[1.41]

Nonetheless, as Laura and Ida's examples suggest, married women were a reality within the workforce. The proportion of married women started to rise during the 1930s as a result of the Depression. With the burden of reduced employment, especially at the vulnerable Sessions, working-class families developed new strategies to survive. In Bristol, men and women who worked during the 1930s recollected that families assumed a buffer against short wages and no work, all pointing to a growing tendency among management to hire more and more married women eager to augment their families' earnings. By the 1930s full-time employment for married women became increasingly a regular part of life for this group; and was desirable because of the good pay. However, the jobs were not permanent; the owners fired women during layoffs before their husbands. Still, this flew in the face of the owners' belief that married women stayed at home.

The reality was, of course, that local families often sent multiple members, whether single or married, into the well-paying clock and watch industry. After graduating from high school in 1937, Dorothy "Dottie" Beaucar became a timekeeper in the watch department at Ingraham, in whose case shop her two brothers also worked, one as a router, the other as a timekeeper. They all contributed their wages and worked in the family garden to raise food for family consumption. Both brothers had paper routes, and one also worked part time in a local grocery store. However, the two men and youngest sister soon married and left home, leaving behind Dottie. Never marrying, she lived at home with her parents and took care of them. After her father died in 1954, she devoted herself to her aged mother and relied on her Ingraham wages as the family's primary support.[1.42]

Not uncommon at all were families with wage earners at both Ingraham and Sessions. Peter Burns, a trucker on assembly of 8-day alarms at Ingraham during the early 1950s, remembered the following scenario concerning himself and three siblings. While Peter was at Ingraham, his sister Eleanor, or "Ellie," worked at Sessions; she started in 1944 as a bezel assembler and remained as a line supervisor until the firm closed in the late 1960s. One of his brothers also worked at Sessions as a trucker from the late 1940s to the early 1950s. Another brother, Louis, who was blind, worked there as an inspector of company-made military fuses after graduating from a trade school for the blind in Wethersfield, CT. At the same time, Peter's family had other relatives employed at Ingraham. His paternal aunt and uncle and the latter's wife all worked at Ingraham. Of the immediate family's wage earners, only Peter's father worked elsewhere; he was an employee at New Departure, where his son joined him in 1953 after leaving Ingraham. No longer at Ingraham, Peter married a Bristol woman whose mother, brother, and maternal uncle all worked there.[1.43]

For the majority of families who either followed a company culture or were more industry oriented, children often entered the labor market with the companies that employed their parents and relatives, a strategy in line with the owners' hopes of creating an efficient workforce. "It was very family-oriented," recalled Anna Calderoni, a former watch inspector, who got her job on watch inspection during the 1930s through her sister doing the same work. "One person in the family worked and they got another person in and there were a lot of people in one family that worked there. And that's how I started." George Power, who became plant superintendent in 1974 after starting as a trucker on 8-day clocks in 1952, also took note of Ingraham's family orientation toward workers. "See, Ingraham's had a history of hiring family," George explained. "[I]t was very common to have a father, mother, sons, daughters, and right on through, if you were a good worker, and your recommendation went good, as far as bringing in your family." That way, "you stayed there a long time, you see. They had long time seniority people there." Employee records prove him right. In 1947 Ingraham had on its payroll 121 employees who had worked between 25 and 50 years each, five between 57 and 60 years, and another five pensioned retirees each with between 50 and 56 years. In 1954 Sessions Clock boasted three retirees with 160 years of work between them and sponsored a Quarter Century Club for roughly 50 workers with between 25 and 60 years of service to the firm.[1.44]

These family ties were all the more powerful because of other facets of company paternalism beyond wages. Advocating what Edward Ingraham called "laissez faire," Ingraham operated with a hands-off dependence on subordinates and their loyalty with the expectation that foremen, supervisors, and workers did their jobs honestly and efficiently in a hierarchical family setting. As "parents," the owners/managers acted whenever possible to grant the needs and requests of productive workers. Kay Laviero, at Ingraham from 1928 until 1952, used this privilege to regularly ask for high-paying jobs. "Every job I had [at Ingraham], I always asked for it. You can't do that today."[1.45]

Such latitude was especially crucial during the Depression when Kay worked to maintain her birth family's household economy. Kay, who was born in Bari, Italy, in 1913 and immigrated with her parents the same year, remembered being pulled out of high school in her second year because her mother was sick. As the oldest child, Kay had to run the family household until her mother got better. Although her mother soon recovered, Kay's father decided she could help support the family financially and secured her a position at Ingraham at age 15. After her parents separated as a result of her father's domestic goods store failing during the Depression, Kay took over as the primary breadwinner for her family. Kay's income allowed her mother to remain at home and raise five sons and one other daughter, all of whom went to high school. As one of her strategies to augment the family income during the early 1930s, Kay made sausage sandwiches with her mother early in the morning and sold them in her department prior to the starting bell. A disapproving Edward Ingraham personally ended Kay's sandwich business, but she used the opportunity to get a better job. "So after [Ingraham] left, my foreman came up to me and said, 'You can't sell sandwiches anymore.' I said, 'Well, you'll have to put me on a better job because I want to make more money.'" When asked if she got a better job, Kay gushed, "I did. I went to work upstairs [to higher-paying watch work]." Empowering workers like Kay further, the owners turned away from the extensive time studies popular at the moment under Taylorism. "We felt that this [Taylorism] was not," remembered former executive vice-president Bob Tetro, "our cup of tea."[1.46]

This empowerment was also prevalent at home, where the majority of single workers lived between 1920 and 1960, especially during social or economic upheaval such as the Depression and World War II. On the one hand, this setup was restrictive; young men and women were socialized to learn to treat money soberly, as the owners no doubt hoped. Living at home often meant contributing wages to the household economy. Men kept a larger portion of their wages than did women; male former workers recalled keeping a few dollars for themselves while women uniformly recounted getting only the loose change from their paychecks. This gender difference was particularly visible in the poorest families. "The whole thing went to my family," Dottie Beaucar recalled. "I didn't even open my envelope." In contrast, Kenneth Webster, who went to work at 16 to help his widowed mother support their family, retained a portion of his pay for himself. As with most women employees, Nelson Spring kept just the change in his envelope—46 cents—when he started at Ingraham as a 14-year-old stock boy in 1918. He received more only after holding a full-time job in his later teens on his way to becoming eventually plant superintendent. Nelson's example suggests that, at least in earlier periods, age mattered as well as gender.[1.47]

On the other hand, young workers found independence. With the pay they did not surrender to the household economy, single men and women turned to consumerism, often along gendered lines. Usually, such spending reinforced on-the-job networks, as Lois (Mastrianni) Cieszynski, Laura Santago's sister and a Sessions worker from 1948 onward, found by going out to eat. "We had a lot of fun and laughter," recalled Lois, "and we used to have an hour's lunch in those days, and we all used to have our lunch together. We used to play cards and we used to go to Johnny's Restaurant [in Forestville] and have our lunch and a few drinks, and that was fun." When off the job, reported interviewees, women most frequently bought clothes or material to make their clothing, often shopping together in groups. A shared identity as fashion consumers in turn spilled back into the workplace, leading by the late 1940s to friendly competition among co-workers at Sessions, as Lois explained:

Question: Was there ever any jealousy [about women] dressing well?

Answer: Oh, yes, there was. That was great. I loved every bit of it. There was a lot of jealousy of the way people dressed. In fact, we used to have clubs and we'd join Dorothy's in Bristol—it was a clothing store—and Miller's, and we used to pay so much a week with the club and everybody would buy their clothes and see who could dress the best. There was a lot of competition.

Women who had extra money left over indulged in amusements like roller skating or dancing at the popular local destinations of Lake Compounce in Bristol or the smaller Paderewski Park in Plainville. Participation in consumer and mass culture was limited and contingent upon what money young workers living at home were allowed to keep. Walantyna Sakowski, the only child of poor Polish immigrants who began work at age 23 on Ingraham's 8-day clock assembly in 1945, remembered having only enough for bus fare.[1.48]

Male interviewees spent their earliest earnings on candy and movies. By their later teens they used their wages on sports like pool, bowling, golf, and group trips to Yankee games, beer, dates and, like their female co-workers, dancing at local parks. Car ownership became common among male interviewees born after 1920. Beginning in the late 1940s, men from this group bought cars, usually secondhand, which they paid for and maintained through their wages while living at home.

Emboldened by their relative independence through wages, young men and women also made additional money on their own by moonlighting. Unlike the earnings from their main job, which were subject to the family claim, the money from this extra work was theirs, and they often spent it accordingly. Jerry Nocera supplemented the two dollars he kept from his paycheck at Sessions in the 1930s with a regular—and grueling—paper route of 350 deliveries a day, as well as cutting hair in the family barbershop. Similarly, Kay Laviero earned money exempt from parental appropriation by performing other jobs:

During my lunch hour I would go down and set their [co-workers'] hair, you know? They would pay me a quarter to do their hair and then sometimes I would also do knitting. My mother and I would knit sweaters. We did a lot of knitting. I would take orders from the girls in the factory. That's how I earned my extra money. So that extra money was mine, not my mother's, but my pay envelope was hers [laughs].

Willing "to do anything to earn as much as I can," Kay likewise worked one weekend day in a small dress shop. The aggregate income compensated Kay for forfeiting her factory wages and the money from selling sandwiches. The sum also allowed her to buy clothes for herself, and sometimes "things" for her younger sisters as well. "That was a luxury for me to spend on my sisters, too."[1.49]

The economic individualism that resulted among sons and daughters who retained a portion of their wages and at times increased their personal earnings with extra jobs often meant the decline of parental authority, implicitly challenging the hierarchy cultivated by the owners. For the youngest workers, such liberation could actually cause guilt. Early in his career, Nelson Spring took more from his earnings than allowed and on his way home bought candy, an act he soon regretted because he had dipped into family money. "I didn't dare to take the candy home . . . so I buried it under a stone and the next day I went [laughs] to the stone to get some candy and it was all full of ants [laughs]." Older wage earners were far more defiant and less apologetic than Nelson. As he got into his later teens, Jerry Nocera decided to keep his entire pay and give his parents what he felt was proper for board. Jerry's decision had far-reaching consequences for his relationship with his Italian-immigrant father:

> I wanted to give him fifteen dollars a week because I was on piecework and I was making more money, but he refused that so I left home. I went to live in Bristol with one of the fellows that worked with me down to Sessions. Then . . . finally . . . [my father] wanted me to come back and my mother called me up and says, 'He'll take the fifteen.' I says, 'No, I'm only going to give him ten now.'

Jerry's father, backed into a corner, acquiesced and took the $10 from his son, a future union steward at Sessions.[1.50]

Daughters likewise undermined parental control. Kay Laviero's father forbade his daughter to date, despite putting her to work at age 15. When she was 18, Kay met a local boxer and the two quickly became romantically involved. Kay hid her romance from her parents, and especially her overprotective father, by seeing her boyfriend only after work when he secretly walked her home. When a disapproving neighbor informed Kay's parents, her father became furious. "My father just kept scolding me," Kay recounted. She responded by confronting her father directly and turned the situation in her favor. "I said, 'Well, if I'm good enough to go back and forth to work, I think I'm entitled to go out with a boy.' You know, really. Oh, he didn't like it at all. He said, 'Do you want to go out with the boy?' 'Yes, I do.' So then he says, 'I want to meet his mother and father.'" Although having the final word, the father did not again challenge his daughter, who later married the boxer.[1.51]

Among most wage earners, dating happened usually between the ages of the late teens and early twenties, with marriage often following. Interviewee accounts are filled with men and women marrying as a result of workplace romances, with marriage ages usually hovering in the early to mid-twenties, regardless of sex, collar status, or year of birth. If not marrying within the workforce, employees found spouses in other local factories, especially New Departure, many of whose men married women from Ingraham. Regardless of workplace, they tried to plan their finances accordingly whenever possible. As low-paid watch inspectors, Al and Rose Calderoni "made only twenty-five dollars between the two of us" at the time of their marriage. The pair nevertheless saved a hundred dollars and paid for their wedding plus a honeymoon to Canada, in part by making a premarital decision to live in the household of the bride's mother upon their return. Higher-paid workers had less of a problem. Kay secured her boxer boyfriend a job at Ingraham a year before their marriage to save money. They pooled their incomes and bought a new car, a 1937 tan Chevrolet. In addition to facilitating travel to and from work, this vehicle allowed them to travel to New York City for their honeymoon and Niagara Falls during Ingraham's annual August two-week vacation.[1.52]

Once deciding to marry, workers were almost always thrown showers, grooms as well as brides, enhancing the family image of the plants. If the intended marriage was among two workers within the plant, a couple often had two showers, one for each person. Two showers, however, were not always the case, and the woman—the traditional recipient of such events—was sometimes not given one. In the instance of Al and Rose, the groom had one while his bride did not. In other examples, brides enjoyed multiple events. Sessions case department superintendent Arthur Brightman and his wife gave his sister, a production worker at the plant, one of the two showers she had before her marriage in November 1912. While this first shower occurred at her brother's home, the bride-to-be's department held a second celebration at work. Sessions' Lois Cieszynski had a nearly identical experience 50 years later in 1962 when marrying the baker who supplied the company cafeteria. "They gave me a nice big shower," Lois exclaimed. "[Longtime worker] Lottie Kerr on Anderson Avenue, she had a big shower for me at her house. They [also] gave me a surprise shower at Sessions in the factory one lunch hour."[1.53]

Most weddings for workers were in fact departmental affairs and invariably reflected factory settings, with co-workers often in attendance. At Lois' celebration, this reality had whimsical results that nevertheless demonstrated the close-knit nature of work and community. After the surprise shower, "then they all came to my wedding," Lois gushed:

> And the morning they all came to my wedding, I think about ten people got stopped for speeding [while traveling from Forestville to Virginia's native Southington where she got married] and the cop finally said, 'Who is this person that's getting married?' and [they responded] 'We just work with her.' So he finally let everybody go because . . . everybody was stopped for speeding.

In the family-like factories, workers shared the thrill (and perhaps anxiety) of the bride and groom.

Following showers, weddings, and—if lucky—honeymoons, newlywed couples returned to wage earning but now work took on new meanings, especially for women. Whereas men aimed for continued wage earning, married women faced the question of whether to keep working or to have children. Confronted with the prevailing emphasis on children, married women frequently made the choice to leave work, at least until children reached school age. Taught from youth to equate womanhood with motherhood, most women made this decision willingly. If wanting both a child and a job, women could be pressured from their husbands, who forced their hand, as was the Ingraham inspector Bertha D. following her marriage in 1940. Bertha liked her work very much, especially since her mother was also an inspector in the same department. However, she also wanted a child. When she expressed this desire, her husband, who worked elsewhere, informed her, "If you want to have a baby, you're going to stay home and raise it." Bertha acquiesced and stayed home to raise her child for the next eight years, before returning to work at another job.[1.54]

Husbands measured their personal worth as men by their ability to meet household demands for cash. In assessing their self-worth as breadwinners, married men were especially likely to compare themselves with their fathers and their wives' positions with their own mothers. The identification with fathers could be traced to childhood, when fathers were likely employed where their sons one day would be. When asked about his childhood hopes for an adult job, former plant superintendent John Denehy recalled, "Well, I thought the same as my father [who] worked in Ingrahams [as a watchmaker]. I figured I wanted to work in Ingrahams, too, when I grew up." John offers a unique case in that he witnessed his father's wage earning capabilities at home. His father, who had charge of the vibrating room, regularly brought home pocket watch hairsprings to vibrate or adjust, the only homework that the company supported by the 1920s. Impressed by the money his father made at home and at work, John determined to go into watchmaking. Other sons, who did not witness their fathers' paid labor so directly, nonetheless styled themselves in similar ways.[1.55]

Adulthood characteristically brought a desire to surpass their fathers' wage earning so as to prove further their manliness. In this regard, husbands frequently linked gender to consumerism and looked toward a typically U.S. goal of male economic individualism: home ownership and the ability to keep their wives gainfully unemployed and in the role of housekeeper. Home ownership became an increasing reality for many workers at Sessions and Ingraham in the post-1945 period with low-cost housing in and around Bristol, paralleling contemporary patterns elsewhere. At Ingraham, workers frequently financed such projects through loans from the employee-run Ingraham Credit Union, founded in 1940. While still in blue collar work, George Power borrowed $3,000 to make a down payment on his first house after his marriage. Others employed different strategies. Eusebe Simard, a machinist at Ingraham, maintained his family's house—and his wife as a homemaker—as a proprietor of furnished rooms during the 1940s and early 1950s.[1.56]

In pursuing economic individualism through wages—and sometimes other paid labor—husbands often allowed domestic tasks and child raising to fall heavily on their wives, opting instead to do "men's" work when home. John Denehy remembered that after he married in 1934 while a watch adjuster, his wife did all the housework and continued to do so despite having four children by 1940. Questioned if he ever helped her with housework in this period, he replied, "Oh, yeah. I was the best husband you can get." When asked to elaborate on the jobs he did, he revealed that he really did not do housework at all, at least not in a traditional sense: "Well, I always had two cars, a couple of junks, and I got to where I could really work on cars, especially in them days." Cars, which were out of doors and associated with men, were John's priority at home. Exceptions existed. Examples include babysitting, occasionally helping with dishes, or maybe vacuuming the home—but the latter chore usually only if the wife did not feel well. However, such activities were indeed exceptions and in no way represented the norm.

In contrast, marriages in which both spouses worked appeared more egalitarian in the allocation of domestic tasks, especially if at the same factory. Interviewees from this group divided housework and child raising more evenly than any other set of married workers interviewed. During the 1950s, Domenic and Sandy Giammateo worked separate shifts at Sessions after the births of their several children. Under this arrangement, Domenic helped to manage the household. "[When] I got onto second shift," Sandy related, ". . . he came home and pitched in. . . . I used to start supper or got supper ready, so then he would pitch in with everything."[1.57] Even in more egalitarian marriages, however, wives performed the majority of housework, unquestionably because custom dictated such labor as woman's work.

Nevertheless, married women in the workforce relied

on various other resources to facilitate their household labor. Sandy counted on her mother, who lived in an upstairs apartment, to babysit her children when she left for work and Domenic had not yet returned. Others relied on their own cunning to entice their husbands into a more equal distribution of work; Kay Laviero convinced her husband in this way. Laid off from Ingraham three months after they married, he later opened a package store in Bristol, which precluded him from going to work at 7 a.m. like Kay. Although confessing he wasn't very domestic, Kay used the situation to her advantage:

> I said, 'I can't get up in the morning and make breakfast.' So he'd get out of bed so I could get dressed to go off to work and he'd have my breakfast ready. Then from then on, it was, 'Oh, you make very good breakfast! I love your breakfast.' From then on, he made breakfast every morning. I conned him into it.

Apparently, this arrangement continued until Kay left Ingraham to help manage her husband's store.[1.58]

Married women also eased the burden of household labor by sharing it with other wives and mothers they met on the job. As did many factories, Ingraham and Sessions operated individual departments on a sex-segregated basis. In these environments, solidarities along lines of gender and class typically developed among employees, facilitated by family and neighborhood bonds among co-workers. In turn, these relationships frequently overcame the impediment of multiethnic origins and forged friendships that extended back into the community; interviewee accounts are filled with stories of U.S.-born workers who became friendly with immigrants and their families both on and off the job. Italian American Philomena "Phil" Carrier got to know well native-born Dottie Beaucar when both worked in the watch department. When Dottie's father died in 1954, friendship inspired Phil to lead a collection for her. With the department superintendent's approval, Phil collected $200, or one dollar from each of 200 men and women employed on watch production, for Dottie's benefit at home.[1.59]

Support networks were particularly prevalent among married women. With their growing presence at both Ingraham and Sessions after the 1930s, they sought each other out with increasing regularity on the job. Motivating them was not only a shared workplace identity but also their roles as wage earners and wives and mothers. This duality created a common pattern of jobholding, which reinforced their mutual identification. Regardless of the job performed, most married women after the 1930s moved in and out of paid work, depending upon the pressures of home and family and abetted by the seasonal nature of industry production. The annual production of clocks at Ingraham and Sessions reached its zenith between fall and spring, and then fell off dramatically in the summer. Both companies also managed the later addition of fuse contracts, likewise cyclical, to coincide with busy periods. Typically, married women, whose often-interrupted work records frequently denied them seniority, were the first to be laid off during summer slack periods.

Many wives and mothers with school-age children actually welcomed summer layoffs. Lillian Rock, a mother of two school-age children in the mid-1950s, enthusiastically recalled being let go when working as an "Ingraham girl" on war work. "I got laid off every summer which was fine. I was with my children." Like other women, Lillian moved easily in and out of work at Ingraham despite a divorce in the mid-1950s. "In the meantime, we divorced because [I] had remarried [in 1958] and when I was called back, I didn't go back because I was expecting my third child. I didn't go back . . . until [that and a fourth child] were both in school." While such maneuverability unquestionably augmented the benefit of wages and made work attractive, it also gave married women a common identity and fostered unique bonds that their unmarried female co-workers did not share. Agnes Moquin, a Sessions worker on both clocks and fuses while a mother of six children in the early 1950s, confirmed this. In recalling her co-workers' habits of association on the job and off, Agnes noted that married women "hung together" while single women had their own groups.[1.60]

In hanging together, married women solidified emotional bonds and social networks and shaped them in response to their position in the sexual division of labor in the home, the fulcrum upon which any participation in paid employment revolved. In this way, married women from ethnically heterogeneous neighborhoods worked together to fulfill such obligations and responsibilities as housecleaning, cooking, child care, laundry, and nursing. Married women interviewed related that hanging together off the job became an integral part of everyday life. Women shared car rides to work, especially from outlying areas. When they planned picnics or even holidays like Christmas, they frequently combined resources and families, opting to have joint outings and celebrations and divide the work rather than do all themselves. At work, they frequently shared their snacks. Kitty Baldaccini used to bring in a jar of peaches that she canned to share with others on Sessions' conveyor line while a woman beside her "brought nuts in and in between doing the clocks we would eat the nuts. Everybody like [sic] would bring in things." In leaner times, women shared whatever they had between them. Rose Calderoni, who went to Ingraham in 1934 on watch inspection, recalled sharing with three or four co-workers with whom she countersunk fuses during World War II. "At that time, everybody got together whatever little we had. Say we bought even a piece of cake or anything and there was four of us. We'd cut it in four pieces so we'd each have a little piece." In contrast, men shared little food of any kind with each other on the job, opting instead to follow the belief of "every man for himself" that characterized other aspects of their work relations.[1.61]

The role of these networks culminated in the willingness of married women to help others in extremely desperate situations. Kitty remembered trying to save an alcoholic co-worker's job during the 1960s because the woman's husband had just had heart surgery and would need her when he came home. The widowed Kitty not only talked to the afflicted worker and got her to enroll in Alcoholics Anonymous, but went to management twice on her behalf and then openly confronted a Sessions stock boy and a local liquor store, two sources of the employee's alcohol supply. Kitty lost her fight on the married woman's behalf because of continued alcoholism, but her time and effort reveal the extent to which networks between married women operated.[1.62]

The Limits of "Family"

Workers could be fiercely loyal to one another. This loyalty undermined the owners' hopes for homogeneity by fostering labor patterns rooted within working-class families that created powerful, gender-specific workforce solidarities that negated the very idea of uniformity. For married women, whose wage earning was frequently interrupted by their roles at home, paid labor could not be completely reconciled with extra-market activity. With their numbers increasing after the 1930s, they demonstrated a willingness to sacrifice for each other at home and in the community. In contrast, male interviewees talked little of helping their co-workers in the physical and emotional ways that wives did and looked foremost to economic individualism and providing their own families with the necessities of life.

This difference reflected larger tensions in the paternalism that both the Sessions and Ingraham families practiced at their plants. To be sure, the variant of paternalism offered by the owners provided workers with numerous benefits; paternalism as practiced in Bristol went beyond the general understanding of this management ethic. Paternalism, Richard Sennett writes, is "the authority of false love"—false because it keeps the worker inferior and dependent.[1.63] At Ingraham and Sessions, paternalist policies empowered workers through steady work and strong wages. Arguably, however, these same measures may have also removed social barriers that fostered worker dependency and incited rebellion. Because class

Figure 1.17. Sessions women on a railroad car behind their factory, late 1940s. Interviewee Sandy Giammateo is at the top.

derives its significance from hierarchically structured access to productive and authoritative resources, workers did not fully identify with owners, and vice versa.

Tensions regularly revealed the underlying distrust that existed between workers and bosses. For example, on the afternoon of April 16, 1923, 45 women in the marine finishing department walked off their jobs after presenting a petition for a 10 percent increase in piecework rates. Although management was in fact in the process of adjusting rates equal to the amount demanded, not all the women agreed about the new piecework system; before presenting the petition, one worker arbitrarily changed its terms from 10 to 15 percent. Management's flat refusal to grant a raise at 15 percent sent the women packing. They returned to their jobs two days later only after resolution of the misunderstanding. These workers did not buy wholly into company paternalism, nor did the owners. As soon as the walkout began, William Ingraham issued a letter to local employers, Sessions Clock included, in which he blacklisted the 37 single and eight married strikers, requesting that they not be hired again in Bristol. William followed a pattern typical of welfare capitalists: treat employees as a loyal family members when they behave and lowly ingrates when they object. Only a few years prior, W. E. Sessions had shown a similar wariness that extended to a general concern about the larger working class in Bristol. During World War I, he supplied Colt automatic revolvers to the Bristol Home Guard, the stay-at-home militia, in part to thwart local organizing activities of the radical Industrial Workers of the World (IWW).[1.64]

Class differences about the meaning of family came to a head in the spring of 1941 when workers at both Ingraham and Sessions Clock won union representation through the CIO's United Electrical Workers (UE). In turning to collective bargaining, employees of both firms openly rejected the owners' politics of protection. As wronged parents, each family let their displeasure be known throughout the workplace. Looking back at the company's management style prior to the unionization campaign, Edward Ingraham reflected, "I supposed our attitude toward employees was paternalistic and we had in general a happy and contented set of employees." The

CIO's victory, however, undermined this idyllic scenario for Ingraham, who confided, "After the Union gained the right to represent our employees—with strong government backing—our attitude changed," adding in a declarative fragment, "No longer paternalistic certainly." The owners, with their immediate subordinates, subtly distanced themselves from their workers by restricting workers such freedoms as asking for favors from bosses, as Kay Laviero had previously enjoyed. At Sessions, Johnny Nocera remembered that after the union came "they hired these efficiency men down at Sessions Clock to make more money for the company." With owners' support, the experts immediately cut back the liberties of the workers in a more systematic fashion at Ingraham by putting wire around each department. "You couldn't go from one side to another. You had to stay in your own department, see. They were like caged in." The Sessionses' feelings manifested themselves fully by June 1942 when they exacted their final revenge by laying off all union members during the conversion process to military production and killing the local.[1.65]

Family owners continued nevertheless to don the veneer of paternalism for the remainder of their tenures. They worked hard to bolster profits and offset costly employee incentive programs, especially at Ingraham where union activism kept management on its toes. Both firms instituted new product lines and sales strategies to remain competitive. They effected their first major change at the end of World War II in 1945 by dropping the production of key-wound 8-day pendulums, which consumers viewed as antiquated. At the Forestville factory, President W. K. Sessions and his son (and namesake), took the additional step of curtailing all spring-driven clockmaking. In the face of growing imports in 1946, Ingraham executed a drastic maneuver by introducing the Sentinel Line of spring-driven and electric timepieces, which the company advertised for the first time directly to consumers as well as to retailers. Sessions followed suit and diversified its offerings by the early 1950s to include miniature switches and industrial timers. Under W. K. Jr., or "Bud," who was named secretary and general manager in 1945, the company even adapted a new engineering division to remain competitive. However, both sets of owners were more backward looking than not and after 1945 still based their manufacturing on 1930s mass production techniques. Harking back to preunion days, they continued long-standing paternalist practices that hinged on a hands-off reliance on older bosses who had invariably risen through the ranks earlier in the century and whose knowledge of mass clockmaking was dated. For the Ingrahams in particular, the old-style managerial faith in the abilities of a skilled department head or foreman had proven well-founded, most recently with the outbreak of war: it had been Ferdinand Fengler, a boss from the tool-and-die room, who designed the first mass-producible antiaircraft fuse, an accomplishment he had achieved by laminating and strengthening regular watch plates. The past success of the owners' policy, however, blinded them after the war from recognizing that the same older bosses upon whom they relied for profits used knowledge that was obsolete against the rapidly evolving technologies embraced by foreign firms. Such a mindset limited the Ingraham and Sessions' ability to compete with their competitors abroad. Ingraham began to use imported Swiss watch movements in 1951 while still making its own, paradoxically believing that the survival of their prized watch line depended on using the very items that so hurt profits.[1.66]

Figure 1.18. Ingraham's executive management in the main office, 1950s. Edward Ingraham, third from left, his brother Dudley, second from right, and Bob Tetro, at Dudley's right.

Eager to evade responsibility, the Ingrahams looked for scapegoats. In 1951 they blamed their sons, then being groomed for management, for not being up to the task of running the company and threatened to bring in outside managers. Edward Ingraham unilaterally blamed the union and his workers. They, he believed, undercut the competitive edge against foreign companies. "Our employees," Ingraham sneered in a sweeping indictment, "did less and less work and were less and less happy and we came more and more to lose our position in the in-

Figure 1.19. General Manager W. K. "Bud" Sessions Jr. from Sessions, *127 Years of Clock Craftsmanship*.

dustry." After profits fell in the mid-1950s, the stubborn elders elected businessman Robert Cooper of Chicago in August 1956 as the first non-family president in the firm's history. While W. K. Sessions Jr. tried to persevere, the unionization of his company in 1956, by the AFL-CIO's International Union of Electrical Workers, proved the breaking point. With the union inside his factory and declining sales, Sessions felt unable to control the fortunes of the family firm and allowed the sale and takeover of the company to an outside corporation in 1957.[1.67]

Figure 1.20. Subassemblers (left) and assemblers (right) at Sessions, ca. 1950. From Sessions, *127 Years of Clock Craftsmanship*.

As symbolized in the different positions taken by the owners and workers, class alienation persisted throughout the paternalist era. Under this reality, the paternalism of the owners could not entirely succeed. Rather, by curbing fundamental liberties of workers while simultaneously providing them with benefits unparalleled in Bristol, welfare capitalism worked against both itself and the men who employed it. Workers identified with the owners but were loyal foremost to each other. The Ingrahams and Sessionses also practiced their own variant of a "negotiated loyalty," likewise based upon the prerogatives of class. When workers challenged management, their employers reacted against them, exhibiting a fundamental distrust that undermined the very ethic upon which the owners based relations with their employees. As a result, the image of a happy family was at its core more myth than fact.

Chapter 2
The Men and Women of the Factories

Carl Kirkby was proud of his many years as a tool and die maker at Ingraham. Because his father was a longtime employee there, he already knew the appeal of being a well-paid, expert toolmaker when he started in 1936. "That's what I always wanted to be," he asserted. Through his father, a highly skilled and respected watchmaker, Carl circumvented the long list of would-be mechanics and gained access to the coveted toolroom. The 16 year old worked hard, surviving the rigors of a lengthy apprenticeship and ultimately mastering his job. As the years passed, he felt he had earned his place among other men in the department, whom he termed "the big boys" of the factory.

Despite his self-confidence, Carl did not behave as a "big boy" with all Ingraham men, especially his bosses. After serving in Europe during World War II, he returned to work as a married man with a new home. With a child on the way, he wanted more money to help defray the costs of starting a family. With hat in hand, he deferentially approached "my big boss" in the toolroom, the respected German-born Louis Burghoff. "I said, 'Mr. Burghoff, I'm expecting an addition to the family in a few days and I just wonder if there's any chance of a little raise. A little more money would help with the hospital bills.'" Deference paid off; Burghoff checked into the matter and a few days later told his subordinate, "Well, Carl, I got you a dime [more an hour], how's that?" Pleased at the outcome, Carl maintained his submissive tone. "I says, 'That's great. I'm very grateful, Mr. Burghoff.'"[2.1]

The expectant father carefully went about his request for a raise, and in doing so endorsed the complex pecking order in place to achieve his aim. Like Carl, local men and women entered Ingraham and Sessions Clock as workers seeking income, most with the hope of making as much money as possible. Inside the factory, they were socialized into hierarchical work environments reflecting a highly stratified wage structure that varied widely according to job, sex, and department, on the one hand, and differing meanings of waged labor on the other. Employees faced multiple divisions of skill levels and types that reflected both the nature of the work performed and customary notions inherent in the industry. Seen as society's breadwinners with a lifetime of wage earning ahead of them, men equated their self-worth with steady work and high wages. They supported the existing power structures as the basis of manly competition with one another and strove to implement hierarchies of their own making to further such rivalry. Their behavior contrasted with women, who derived their identities through the 1950s not as aspiring breadwinners but rather from their societal roles as wives and mothers. Like men, women took pride in their jobs in the effort to earn wages. However, because most expected ultimately to leave the factory to find womanhood at home and in the community, women often opted to follow group patterns at work.

Inside the Factories

Factory life for workers revolved around the many departmental divisions within the companies. Both firms made all parts from screws, plates, and wheels to bezels and the wood or metal cases that housed finished movements. A 30-hour marine clock made at Ingraham in 1938 required 133 parts and 291 operations to put together. At the same time, an Ingraham pocket watch comprised 84 parts and 340 jobs to assemble and a company-made wristwatch, 81 pieces and 298 operations. Altogether, the finely divided labor of both spring-driven and electric timepieces allowed Ingraham to make 100,000 clock and watch movements a day by 1938. The firm's case shop used similar processes in making elaborate wooden clock cases and radio cabinets, namely, for Emerson Radio and Phonograph Corporation.[2.2]

Although much smaller, Sessions Clock also used a complex division of labor. By the middle of the 1920s, each firm numbered 27 departments through which they subdivided the numerous jobs related to the production and assembly of timepiece parts. Within individual departments, jobs were skill based. For example, although a man's introductory position in Ingraham's watch department as a trucker was easy to fill, another male job like watch adjusting—the highest paying in the department—took years to master and required previous experience with watch work. Certain jobs were simply not open to new employees. While they often had family and friends who got them into Ingraham or Sessions, new workers did not always get the same jobs as their sponsors held.

Complicating the division of labor further within each department was the segregation of work by sex, in line with the gendering of jobs. Some departments were almost exclusively single sex. For example, at Ingraham in 1927 the movement shop's radium department functioned solely around the painting of dials with luminous paint, a job that employed only women; a woman even held the foreman's position until the plant closed the department in the 1930s. At the same time, the automatic screw machine and machine (or tool) departments, also in the movement shop, relied on jobs seen as men's alone. The two divisions employed only one woman each as opposed to 43 and 39 men, respectively. Likewise, the firm's case shop, which focused on heavy woodworking machines, offered mostly male-type positions (Table 2.1).

Most departments, however, were more evenly integrated. In keeping with the industrial practice of segregating the sexes at work, Ingraham and Sessions seldom allowed women and men to work side by side; instead, they were partitioned off in groups within each department. Even in departments relatively evenly divided by sex, the same-sex ratios persisted over time. Ingraham's watch finishing division, with 78 men and 97 women in 1927, maintained roughly the same breakdown by sex well into the 1950s. In this department, women, who performed all assembly jobs (e.g., pallet fitting, pinning in a balance wheel, and vibrating a hairspring) sat in separate groups from men who did adjusting and repairs.[2.3]

Table 2.1
Age Distribution of Male Employees in the E. Ingraham Company, March 7, 1940

Age Group	All Workers	Men	Years of Service <5	5-10	11-15	16-20	21-25	26-30	31-40	41-50	Over 50
16-18	15	11	11								
19-25	540	252	174	76	2						
26-35	862	416	85	175	144	12					
36-40	269	130	23	43	44	18	2				
41-45	237	138	26	37	40	27	6	2			
46-50	168	112	17	39	37	7	8	4			
51-55	125	87	9	26	21	16	8	2	5		
56-60	82	63	8	23	11	5	10	1	4	1	
61-65	39	35	1	8	14	3	5	2		2	
Over 65	50	46	2	7	14	13	1	1		4	4
Total	2387	1290	356	434	327	101	40	12	9	7	4

Source: E. Ingraham Company, "Age Groups in the E. Ingraham Employ March 7, 1940," Ingraham Company Papers.

As lifelong wage-earners, men expected to amass years of service as the quantitative breakdown of male employees at Ingraham by age and years of service in 1940 demonstrates.

Men's Skilled Jobs

A wide spectrum of jobs was open to men, with only a small number traditionally classified as skilled. These skilled jobs centered in the movement shop and included toolmakers and screw machine operators and, in Ingraham's watchmaking departments, watch adjusters and repairers, whose pay in 1940 ranked close behind that of foremen. Unskilled men dreamed about one day filling their ranks. George Power aspired to adjusting and repairing when he first started at Ingraham in 1952 as a trucker, among the lowest paid jobs in the plant. He recalled:

> [At] Ingraham's, see your adjusters make good money, piecework. See, . . . everybody that went to Ingraham's started [by] trucking, . . . and their goal was to get up to be an adjuster and earn good money and then they'd be able to be a repairer and that was job security. I mean you had it made. . . . [Trucking] wasn't your goal. This was your 'apprenticeship,' your training period. Do a good job there and you work into adjusting, and that's where the money was.

George, in fact, surpassed his goal, eventually becoming plant production superintendent in 1974.[2.4]

Not adjusters but foremen were, of course, at the height of the pay scale and were also promoted through the ranks. "The foreman in most cases," explained former executive Bob Tetro,

> had probably started in that department as a laborer or a worker and then became a setup man then became an assistant foreman and a foreman. He knew that department. The foreman of the plating department might come in from some other company where he had been in charge of the plating department, but in the main all of the foremen had grown in their department and knew their business.

Bob understood also that the paternalist bent at Ingraham went beyond simply employing foremen in this way:

> Elof Carlson, the [movement shop] superintendent, had originally started in a subassembly department, became a setup man, became an assistant, became a foreman and when the then superintendent retired, he was made superintendent. So the whole organization was geared to people who had learned the business in the company.

So prominent was such internal mobility that Bob concluded that "the strength of the company was in their

Figure 2.1. Retirement party for longtime Ingraham foreman James Kearns (front row, third from right), 1950s. Watch department head George Tetro (and father of Bob) is beside him, second from right, followed by Nelson Spring. Behind Spring is John Denehy.

[foremen's] knowledge of their functioning department."[2.5]

Foremen's pay reflected the skills that the owners/managers expected them to have. These positions were exceedingly hard to get and open to only a select group of men with years of experience. During the 1930s foremen comprised only 1 percent of employment at Ingraham, and because of the smaller size of the similarly numerous departments at Sessions, 5 percent there.[2.6] Nonetheless, men could compete for the other more numerous jobs of tool and die making, screw manufacture, and watch adjusting/repairing. Practitioners of these tasks jealously guarded entry to their departments, often limiting access to family members. These departments, like others, operated in paternalist fashion from generation to generation as men learned their trade by observing other males with skill and often passing that skill from father to son. Skilled departments elaborated the most pronounced hierarchies in the factories to limit entry. In turn, men in those departments prized the hierarchies as the most tangible ladders of self-worth in the plants. Whether entering a skilled department because of male kin or friends, young men began a process in which advancement lay in a patriarchal combination of deference to, and patronage by, their superiors. The rigidity of this pattern reveals how deeply imbedded patriarchal and family metaphors were within work and industry culture.

Watch adjusting and repairing offered the highest blue-collar wage at Ingraham. Requiring a year to master, the job averaged almost a dollar per hour in 1940, although company president Edward Ingraham later estimated that men could make $2.50 on piece rates. To perform the work, related former adjuster John Denehy "you'd get the watch and you would regulate it and adjust the watch so it would be proper." Mastering the job meant to reach a production of 30 pocket watches per hour or 10-12 wristwatches in the same time. The difference in number reflected a hierarchy in itself based on skill. The smaller the size of the watch, the more skill and labor were required, making wristwatches the higher paying. If skilled enough, John reasoned, you "could graduate to a wristwatch, which was smaller and tougher to adjust than the pocket watch. The pocket watch compared to the wristwatch was like a clock compared to a watch. It was a nicer job, the wristwatch was." John, who was interested in making money, defined "nicer" as higher paying.[2.7]

For the most skilled adjusters, work on wristwatches led to repairing proper. Repairing included both "inside" work—watches that adjusting could not correct—and "outside" repairs to damaged watches that customers shipped back. Repairmen had little margin for error; they were responsible for "bad motions" in movements (which caused the timepieces to function improperly) and "stoppers" (watches that wouldn't work at all) and had to fix affected watches for free. "There was really a control," John pointed out, "because your name went right with the watch and if it wasn't right you got it back."[2.8] In the hierarchy of adjusting, repairers were men among men in part because of the demands of their work and because, in the inside repairs, they had direct jurisdiction over any inferior work produced by other adjusters below them.

Because of the interplay between skill and pay, watch repairing—or the more general field of adjusting—was highly sought-after work by the start of the 1930s, especially because apprenticeships lasted only a year, compared to the several years, for example, that skilled tool and die makers had to endure. Watchmaking was also expanding because of a sharp rise in demand during the Depression for Ingraham's cheap, nonjeweled watches. "What happened," remembered John, "[was that] business [in the early 1930s] picked up. Adjusting was the job at the time that everybody wanted, and so they needed adjusters."[2.9]

Access to adjusting, however, was limited. "They had the best men in the watch room, too," he boasted, "because it paid the most." As such, adjusters guarded entry to their craft by the 1930s. As with other lines of skilled work, deference and patronage were the entry tickets. Gaining entry to adjusting, however, was impossible without having first done other work on watches and having a family member who worked as an adjuster to arrange a post. Candidates for adjusting had to know the workings of a watch well before advancing. Going into the watch department at age 18 in 1932, John learned adjusting from his father, foreman of the department's vibrating room. Still, he did not take on the job for pay until after first learning watchwork as a trucker responsible for putting watches "on the run," or timing them for accuracy. After "about a year" in this position, John, armed with knowledge about adjusting from his father, was ready for the next level. He also understood the manly status he achieved on his new job, particularly in the context of the Depression. "When I was on adjusting, I was making forty dollars a week, which was unbelievable. That was great money in '33 and '34." He moved through the hierarchy of tasks from adjusting lesser-paying pocket watches to the smaller and more-profitable wristwatches, reaching the latter after "possibly, oh, maybe two or three years" of pocket watchwork. John left Ingraham temporarily in 1940 for war work at the U.S. Time Corporation in nearby Waterbury, but returned in 1945. He then continued his climb with a job as a watch repairman on outside repairs, ultimately becoming the factory's production superintendent in 1954 before retiring 20 years later.[2.10]

Reflected in John's ascendancy, vertical mobility also extended to screw production and tool and die making. Such advancement occurred most prominently among the roughly 80 tool and die makers at Ingraham by 1940 and their ten or so equivalents at Sessions. These men worked in departments where long apprenticeships were mandatory, and new men were constantly and aggressively judged by the experienced and often much older mechanics who labored beside them. In turn, the tool and die rooms of the two companies were the most hierarchical of any of the departments employing skilled labor.

The tool and die departments were the backbones of mass production; they provided the dies from which clock and watch parts were blanked. Given the primacy of its function, working men associated with this department in any way talked with pride of both the room and its employees, especially at Ingraham, whose mammoth example sustained the company's growth for decades through both open dies and sub dies; sub dies were compound work and necessitated greater tolerances.

Because the work required years of experience, Ingraham maintained a department of generally older workers. This policy gave this room the distinction of having the oldest age group in the plant. Interviewees always talked of older employees as the best toolmakers, often in awe of the abilities that they ascribed them. Because of the singular importance of tool and die making, Ingraham was lenient with any bad habits displayed by bosses in this division. Carl Kirkby illustrated this point with the example of an early boss named Mr. Osgood:

> Osgood was one of my bosses. At that time he had to be seventy, long past his retirement, but he was so good that they just wouldn't let him go. He was an alcoholic. He was an old man. His wife had died. He lived with his two daughters in Bristol. He'd come in drunk. Monday or Tuesday he'd come in drunk and they'd fire him. Next Friday they'd have problems in the toolroom and they'd call him and ask him if he would come back Monday, so he'd be back on Monday. He might work for another week or two and they'd fire him again and he'd come back again. This is a true story. That's how much they appreciated that one man.[2.11]

Very simply, older bosses' mastery over the trade guaranteed that they had their way.

Having their way also allowed bosses in the toolroom to dictate the makeup of their department. In the complex webs of family and kin relations that underlay working conditions, these men, like watch adjusters, turned to their own in choosing their subordinates. In Ingraham's toolroom, German Americans had by the 1930s developed a hold on the departmental leadership and favored hiring members of their ethnic community. Consequently, men with names like Edward Stohl, Herman Rudberg, and Arthur Reckert dominated both the toolroom and its offshoot, the engineering department. For non-German neophytes, though, the ethnic concentration could be alienating. "Bill V.," an anonymous native-born interviewee, started at Ingraham winding hairsprings in 1947 and after a short time transferred to the drafting room. There, he felt isolated among the German Americans. "Now I'm not going to get into an ethnic thing," complained Bill, "but they seemed to stay with their own group in my opinion. And anybody who was not German did not really [fit in]." Non-German workers interested in toolmaking and engineering usually needed a powerful voice to secure them a position. Native-born of English ancestry, Carl went into the toolroom only because his father was an expert watchmaker with a long service record at the company and used his standing to get his son the opportunity.[2.12]

Once in the toolroom, patronage from an older man was essential for success, particularly for non-Germans like Bill who enjoyed the support of the chief engineer, Ferdinand Fengler. After a few years of drafting, Bill had not yet overcome the ethnic otherness he felt and, as a result, disliked his job. However, he had in the interim developed a patron in Fengler who "took a liking to me." Fengler opened the door for the young man to begin a "journeymanship" in the early 1950s, which led to Bill's becoming a master model maker at the firm, a job that he readily identified with. Fengler's patronage had another, even more immediate benefit: Bill no longer felt any ethnic alienation, recalling now "I didn't feel like an outsider."[2.13]

Patronage provided other forms of protection for young men entering the trade, who otherwise would have had limited power in the department's rigid hierarchy. After his father got him a job, Carl won the favor of Charles Stotz. Widely respected, Stotz had emigrated from Germany with his family at age eight in 1889 and worked in the department from 1907 until his retirement in 1954. He was, remembered Carl, "one old German guy" who "took a liking for me." Stotz's patronage was critical when, early in Carl's apprenticeship, an older man broke a machine and tried to blame the teenager:

> This fellow old enough to be my father was going to let me take the rap for that machine. I was waiting for a machine . . . and when the man finally got through with it, I went to use it. Then when I got through with it, this fellow came back . . . to use the machine. He says, 'What did you do to the machine?' I said, 'What do you mean, what did I do?' He says, 'Well you wrecked this part of it.' I said, 'I didn't wreck that part of it. That's the way I got it. I didn't notice the difference.' He says, 'No, you wrecked the part.'

Carl remembered that because of his lowly status he felt he had no choice but to defer to the man. However, Stotz witnessed the whole event and informed the young man that he would not let the older man shift blame him to him.[2.14]

Deference was also necessary for success. Submission to one's seasoned, skilled elders allowed new workers to learn their trade and slowly rise in the esteem of the older men upon whose favor their advancement rested. If an apprentice outstepped his position, he suffered accordingly, as Carl experienced:

> [O]ne day I was filing a piece of work. . . . I was an apprentice. [Mr. Osgood]'d sit and watch me by the

hour.... After a while it gets to you because you wonder, 'What am I doing wrong?'... So, finally I rested the file on the bench. I said, 'Mr. Osgood is there something wrong?' Well, he just looks at me and he says, 'What in the hell are you doing?' I says, 'I'm filing a piece of steel.' He says, 'Whoever taught you to file that way?' and I said, 'Who the hell wants to know?' Now this is my boss. This is my boss. I was just a kid.

What happened next put the apprentice in his place: Osgood took him to the department head, who kindly but firmly reprimanded the young worker:

Figure 2.2. Brown and Sharpe screw machines at Ingraham, mid-1960s.

'You know, I'm not even going to ask what was wrong. I can imagine it maybe wasn't all your fault. All I can tell you, you're working for the best mechanic in Bristol. Don't let anybody ever tell you different. He's the best mechanic in Bristol. All I can say to you is try to get along with him for two or three years, your apprenticeship time because he can teach you a hell of a lot.'

Carl acquiesced, deferred to Osgood, and finished his apprenticeship without incident.[2.15]

By bowing to elders and securing their favor, an apprentice in the toolroom advanced toward a position that would bring skill, respect, and the ultimate goal of higher wages. In the tool and die department, workers metaphorically grew into "men" as they rose, by patronage and deference, through the ranks. This pattern was replicated in the next most hierarchical departments at Ingraham and Sessions, the automatic screw machine departments. These divisions produced the small, but essential, screws for clock and watch movements. Ingraham had the far larger screw department, housed in its own building, whereas Sessions confined its machines to a one-floor operation. In the mass production technology of the late 1930s, screw machine operation ranked as highly skilled because of the knowledge needed to maintain and run the intricate mechanisms that automatically made screws from small bits of wire and because of the time necessary to acquire this insight through apprenticeships. As Bob Tetro explained of Ingraham's large screw room:

... the screw machine department again was a trade apprenticeship.... We had the largest Brown and Sharpe department east of the Mississippi, with 250-500 Brown and Sharpe screw machines, maybe one hundred Bannatyne, so forth and so on. These machines were run by screw machine operators who were highly paid because they had served an apprenticeship. They were craftsmen in their day and age.[2.16]

As in the toolroom, young men entering the screw department earned their place only after serving their apprenticeships through deference and patronage. Likewise, the department employed many older men who won respect through years of experience; Howard Sparks, born in 1877, went to Ingraham in 1913 and worked as a setup man on screw machines before his retirement 40 years later in 1954. Furthering the prestige of seasoned men was the ability to withstand and function in the almost deafening sound of the machines. "When you went into [Ingraham's room], you wouldn't expect to hear a thing in there because it was so noisy," draftsman August "Augie" Erling claimed, "... and at that time they never did anything like ear protection or anything of that sort. So a lot of those fellows probably had trouble hearing when they got in their older years." That danger made them all the more worthy of respect, which Augie freely gave them.[2.17]

Fragmenting the line of vertical mobility in this department was the fact that certain machines required more skill to operate than others, and as a result, each type of machine also necessitated its own particular chain of advancement and pay. Brown and Sharpe machines required the most skill and paid the most, followed in order by Hartford machines, Davenports, and Bannatynes.[2.18] Work hierarchies developed around each of these four types, thus quadrisecting workplace relations at Ingraham. Variations of patronage and deference, as well as group solidarity, likewise jelled along these four separate lines.

Unskilled Male Labor and Its Meaning

In contrast to skilled labor stood unskilled work, subdivided jobs that required little training and that employed the majority of men at both factories. Because of their nonskilled nature, jobs like inspection of clocks and watches in the movement shops primarily afforded the opportunity for horizontal mobility among the men who performed them. Nevertheless, unskilled male workers invented the gradations of self-worth that characterized their skilled counterparts and proved their willingness to create hierarchy even when the nature of work did not produce it.

The case shops, where an almost total emphasis on machine production by the 1930s created collectively the lowest paying jobs in the plants, were the locus of some of the least skilled jobs. So deskilled, in fact, was the work that Ingraham's case shop employed 19 of its total workforce of 284 in 1937 as miscellaneous help, who could shift from job to job when needed.[2.19] For men, most of

Figure 2.3. Working on black enameled wood cases in Ingraham's Case Shop, 1914.

whom were Italian born, such a reality was particularly harsh. Trained in a family tradition of highly skilled woodworking, the immigrant workers equated their masculinity with the ability to make finely crafted woodwork and saw themselves as artists whose individual competence was essential to their identities. Low pay and unsteady work during the Depression denied them the artistic ability to take the time they felt necessary to execute effectively their woodwork. Although skilled jobs could not be "industrialized," those that could led to these men forming the nucleus of support for union campaigns that the United Electrical Workers (UE) conducted at the plants in the early 1940s. Successful unionization of both workforces by the UE in 1941 in fact owed much of its success, especially at Ingraham, to dissatisfaction in the case shops as the next chapter shows.

Despite, or because of such work circumstances, men tried to outperform one another and show themselves to be men. Specifically, they targeted pace and quantity, intertwined aspects of unskilled jobs that were the workers' own purview, and rated their manly status on their ability to work fast. "The people gotta get paid according to their ability," explained Oreste DePascale, an Italian immigrant who rose from an Ingraham belt sander in 1928 to foreman of the cabinet and sanding department six years later. Oreste argued for piece rates against prevailing day rates for the particularly unskilled labor of his division, reasoning that "you make'a more and you get paid for more. The other guy who make less get paid for less. It's so simple." Oreste counted himself as a fast, efficient worker, a trait that he believed underlay his rapid advancement. As a foreman, he reserved the better jobs for employees whom he felt approximated his abilities to outperform others in meeting quotas.[2.20]

The idea of the fast worker is, of course, suspect, especially coming from a foreman, because it threatens worker autonomy. Paradoxically, however, male pieceworkers in the case shops tried to work as fast as possible against one another, not to please the boss but rather as an easy way of proving their self-worth in their departments. Cabinetmakers, or assemblers—the most numerous employees in Ingraham's case shop—were one such group. Italian-born Tommy DiSabato, a former cabinetmaker at Ingraham (and Oreste's cousin), boasted, "Everything was piecework. The more you made, the more money you got." Tommy counted himself as one who could make "more," and thus manly. Jerry Nocera, a sprayer at Sessions during the 1930s and early 1940s, hinted at a similar behavior thereby admitting that in the spray room it was "every man for himself."[2.21]

Augmenting the competition were differing prices for the same tasks. When asked how sprayers as unskilled workers were picked for the higher-paying jobs on large clocks, Jerry explained:

> It was up to the individual themselves because if he was a good sprayer, he got the job, but if he was a little on the off-color, you know what I mean? You had to know what the heck you're doing when you're spraying because if you hesitate a minute, you're going to get a lot of paint in one spot or a lot of the clear lacquer or the colored lacquer, regardless of what you're spraying, see?

In Jerry's estimation, only men who knew what they were doing qualified for the highest-paying jobs. His testimony here shows spraying to have in fact been skilled even if not valued as such and suggests that the definition of skill is more of combined cultural assessments with evaluations of the work than the job itself.

Indeed, Jerry reveals a central quality of unskilled labor for case shop workers: the recognition among themselves of skills that outsiders did not legitimize, but that allowed for the infusion of degrees of pride and self-worth, paralleling the skilled jobs of factories. While men on unskilled jobs made less than others on higher-skilled labor, they compensated for lower wages by reinventing their work and validating its skill. Jerry discussed his own job as though it had an apprenticeship and implied he be-

Figure 2.4. Sessions yard crew, late 1940s. Dom Giammateo, far right.

came an expert who was not only able to mix and use all types of color spray but also the more difficult lacquers:

> And then . . . the longer I stood there, the more I taked up the knowledge of spraying, and then from there I went to spraying all kinds of colored clocks. We had to mix our own lacquer. We had a lacquer room, go out there and mix it up in whatever color we used to mix, you know. . . . I done that for a few years and I moved from downstairs in the colored lacquer to the clear lacquer upstairs and that was all clear.

He also took pride in the fact that spraying was among the final jobs on a clock case and therefore an important job. He noted that "when we got through with them, they were sent to the rubbing room, the so-called rubber, and they rub them down, you know, and give them that gloss. Then they come back for the final finish. . . . [And] that was it." Skill here was a combination of aesthetics and place in the production process. Messing up the final steps risked ruining a costly case, giving sprayers the right to view their labor as skilled.[2.22]

By maintaining and reproducing the graded identities common to skilled jobs, unskilled men in the case shops vied against each other. They carefully protected their self-images and judged others to reinforce their own self-worth. This practice stood in contrast to the work relations of women at Ingraham and Sessions. Most women also worked in nonskilled work groups that were likewise single sex. Unlike their male counterparts, however, they commonly maintained the nonhierarchical character inherent to such groups. This simple but important distinction underscores the arbitrary nature of the cultural construction of skill. The fine finishing in which Johnny invested such manly pride could well have been spun as women's work, but it wasn't. If this job had indeed been defined as women's work, men like Johnny would have dismissed it as below them. As men's work, however, it allowed males to shape hierarchies that were foreign to women.

Women and Work

Because of industry sex-typing, working women dominated the "delicate" jobs of subassembly, assembly, and finishing clock and watch movements at the plants. Some women did accumulate long service records. However, the majority worked when young and/or single. Most left at marriage or at the latest upon childbirth—in line with society's expectations for them and in contrast to men whose employment rates by age were higher, as 1940 statistics from Ingraham show (Table 2.2). Whether single, married, or mothers, working women found a variety of jobs that appealed to their wage-earning aspirations and paid a wide range of rates. Although labeled unskilled—or at best semiskilled—by management, women's positions were actually priced according to experience. Ingraham's hourly rates for women increased noticeably from the late 1920s onward, making the firm the highest-paying local employer of women by the mid-1930s. Sessions, although it did not follow Ingraham's dramatic strides, likewise attracted women because of its position as Forestville's chief employer and the many jobs available there (Table 2.2).

Among women at both factories, preferences emerged for particular tasks. The wide range of prices on jobs sex-typed female was a factor in determining job desirability, with higher-paying work being the most appealing, as in the case of men's work. So also was a woman's personal regard for the type of job and her ability to get out production. John Denehy remarked that his wife, who worked at Ingraham while their four children were still young, liked finishing (putting dials and faces on clocks) and especially working on the Broadcast model, because she happened to be good at piecework. "That was a radium dial 8-day alarm clock. . . . [It] was all piecework and this particular one . . . was a good one to make a dollar on."[2.23] John also believed watchmaking to be desirable for women, arguing his point by linking the peculiar associations of piecework at Ingraham to the nature of the work as keys to its suitability. "I'll say that 90 percent of Ingraham's was piecework all the way through the factory, including the women." This emphasis on piecework, John exclaimed, fit well with the character of the watch department. "It was really a women's shop. In those particular days . . . women seemed to fit into the way the watch was put together and so forth. So it was really a great place for a woman to go to work, plus it was piecework." Of course, the women individually had to decide whether they liked piecework.[2.24]

Women in the watch department outnumbered men. In their roles as assemblers, women in this department put the mainspring and wheels together between the movement plates. They likewise performed the harder—and higher-paying—tasks of pallet fitting, pinning in a balance, and opening and vibrating the hairspring before movements went to male adjusters. As with adjusting, the pocket watch was easier to assemble and paid less than wristwatches. Among the highest-paying jobs for women in the department was work on the smaller ladies' wristwatches, in which assembly, pallet assembly, pallet fitting, and pinning in the balance took about three times as long as similar operations on pocket watches. The department, in fact, offered among the highest wages of all women's jobs at Ingraham during the Depression. "Actually the '30s were great . . . [because] anybody who wanted to make a dollar went to Ingraham's," boasted John.[2.25]

Women themselves preferred watchwork to other types of labor. As a reward for being the fastest assembler of 8-day clocks at Ingraham in the mid-1930s, Phil Carrier received her choice of department and the right to take a coworker with her. Phil chose the watch department and took her friend Lil', who couldn't believe this was happening. Summing up the apparent feeling of many women, Lil exclaimed, "Do you know that's the best department in the whole factory?" The deal was

Table 2.2
Age Distribution of Female Employees in the E. Ingraham Company, March 7, 1940

Age Group	All workers	Women	<5	5-10	11-15	16-20	21-25	26-30	31-40	41-50	Over 50
16-18	15	4	4								
19-25	540	288	199	85	4						
26-35	862	446	68	178	157	43					
36-40	269	139	29	52	33	20	5				
41-45	237	99	24	37	23	11	4				
46-50	168	56	3	22	19	7	5				
51-55	125	38	4	13	14	5	2				
55-60	82	19	1	6	4	7	1				
61-65	39	4		1	1	1	1				
Over 65	50	4		1	1	1				1	
Total	2387	1097	332	395	256	95	18			1	

Source: E. Ingraham Company, "Age Groups in the E. Ingraham Employ March 7, 1940," Ingraham Company Papers.
While some women did accumulate long service records in the local clock and watch industry, most of those employed at Ingraham in 1940 did not expect to follow lifelong patterns of employment as men hoped to do.

sweetened by the fact that both Phil and Lil had brothers working on watches. So pleased was Phil with her choice, especially with regard to the piece rates she made, that she remained in the department until Ingraham ceased watchmaking in the late 1950s.[2.26]

Kay Laviero, who started at age 15 in the case shop in 1928 cleaning glasses with a brush and was quickly attracted to the watch department, acted similarly. She began her career in the department on watch finishing, a task she did throughout the 1930s. This job focused on fine-tuning the balance wheel:

> . . . I worked taking care of a balance wheel. A balance wheel was small and you put that on a little small steel block and you had to see that [on] that balance wheel that the heavy part went to the bottom. If the heavy part would go to the bottom, you had a tiny press and you'd take little pieces of metal off and you had to touch it and make that wheel go round, uniformly round.

Having mastered this task during the 1930s, Kay was ready for Ingraham's fuse-making program during World War II as an inspector of fuse movement plates. "I worked on a round disk," Kay recalled. "That's what it was, a plate for a time bomb. If there was like a spur, you had to smooth it out. Everything had to be real perfect, smooth edges. All smooth it had to be. You had to inspect every one of those." Kay was pleased with the money she made during the war. So, too, were Sandy Giammateo and other Sessions women who went to Ingraham before going back at war's end. "Well, there was better money there in Ingraham at the time and I followed my girlfriends," Sandy said "We went together, and believe me, we did make . . . quite a bit of [money.]"[2.27]

Kay returned to balance wheel work after participating in Ingraham's war production program during World War II. Her new job, which she held until leaving in the 1950s, involved even more intricate work. "I was working on the balance wheel [again]," Kay recalled, "and . . . the balance wheel is like the point of a needle and I had to work with a magnifying glass and I had to see that that point was perfect, otherwise it would be a reject." Her accumulated experience attracted the attention of the forewoman, who sat next to her and made Kay an assistant forewoman. "I became assistant foreman so when she'd take off for two weeks' vacation, I would run the place for her."[2.28]

Presswork, which was another piece rate job, also attracted women who could master and exceed the quotas set by management. "Everybody wanted piecework," pointed out Rose Calderoni, a former pressworker, "if they were fast." For this reason, Rose recalled press operation as desirable for her coworkers, but not for her because she was "slow," thus showing how problematic piecework really was. To increase their speed and make more money, many pressworkers developed their own strategies of motion. Such motions, although bringing more money, could cause confusion among coworkers. Lillian Rock related a tale told by her husband when he worked there in 1947:

> He tells the story of women pieceworkers that just

Figure 2.5. Sessions women and driver. Dom's wife Sandy is at right.

were so into their job and the money that they were going to make, but they worked so fast that he thought one of them was having a fit. He went and told the boss that there was a lady down there that was sick. She was making awful funny movements. As it was, the lady was just doing her job the way she did it.[2.29]

When making money, some women felt "anything goes."

In contrast, women uniformly disliked one job at Ingraham: radium dial painting. The company formed its radium department by 1920 in response to the consumer demand for the luminous dials made popular by soldiers during World War I. While the department had a maximum capacity of 35 women, growing health concerns over radium and shrinking public demand by the late 1920s resulted in the employment of only 11 women by 1927, all of whom worked full time. Although, or because, work was steady, the job was not popular. Evidence exists to show that the women themselves knew they were suffering from an industrial disease even before 1927. Moreover, in a morbid twist on the limits of paternalism, radium poisoning resulted in the eventual death of at least one dial painter, Helen Lopotoski, who worked at Ingraham under her maiden name of Miller.[2.30]

So commonly acknowledged was the danger inherent in radium painting by the latter 1920s that prospective employees protested against being placed on such work. Phil Carrier went to the department as a 15 year old after her first job on clocks fell through. "I'm not going in this room!" she scolded the superintendent when he first suggested the job. The older man, however, prevailed, convincing her to try the work for two weeks. The department's second boss herself thought Phil shouldn't be there. She informed Phil, "Well, I'll give you a tip: . . . Do lousy work. . . Instead of painting the numbers, make a slip and go paint down." Although Phil at first resisted, the woman remained adamant: "Well, you're going to do it. You want to get out of this place? . . . Well, do what I was telling you to do." Despite detesting the job, Phil found that nothing, in fact, got her out of the department; Mrs. Marden, the forewoman, refused to let her go because new workers were so hard to find. The boss tried to make the work bearable, reading the paper out loud as historian Patricia Cooper found skilled cigar makers did. Marden's peculiar preoccupation with the obituaries, however, failed to assuage worker worries and caused Phil to "get more nervous in that department," where she remained with no ill effects until it folded in the early 1930s.[2.31]

Working women at Sessions Clock also rated their jobs. As at Ingraham, Sessions women did assembly and worked on presses, doing what Domenick Dellario, a former office worker, called "crude" production. Differences, however, existed. Most of Sessions' presses were kick-operated in contrast to Ingraham, which used both foot and hand variants. Moreover, Sessions had a low ratio of presses for

Figure 2.6. Painting dials in the Radium Department at Ingraham, circa 1927. E. INGRAHAM COMPANY PAPERS, ARCHIVES AND SPECIAL COLLECTIONS, UNIVERSITY OF CONNECTICUT AT STORRS.

any particular subassembly job on clocks, sometimes on a one-to-one basis, whereas Ingraham could fill a room with presses that all did the same job. As well, clock assembly at Sessions centered on two conveyor lines, "A" and "B." All clocks, regardless of type, were put together on these lines, from movement assembly to attaching movements and bezels to cases. Ordinarily, the women on the conveyors worked on one particular model until the run was complete and the line was converted to make another clock, although recollections of multiple model production do exist. Each woman had a specific task, as Kitty Baldaccini, on the conveyor from the early 1940s until leaving the company in the late 1960s, remembered, pointing out that "when you work on the conveyor, you had one particular job."[2.32]

On its conveyor lines, Sessions paid women a flat rate, with the supervisor of each line getting 10 percent above her subordinates by the 1960s. Despite the uniformity of wages, women did find one job harder than another. Kitty thought work on Sessions' popular Westminster chime was difficult. "I had to put the crystal in the sash and I had to adjust the hands," she reminisced. "And, of course," Kitty continued, "that was one of the hardest jobs on the conveyor" because "first you had to fix the hands and . . . be sure that they're not touching and all that." Then, she complained, "You have to wipe the glass and the sash." Generally, however, conveyor line work was the most enviable among women at Sessions. Lois Cieszynski, who started at age 16, spent a 20-year career doing the various jobs of conveyor assembly, press operation, and packing clocks. When asked which work she liked best, Lois named the conveyor. "Well, the conveyor I liked it because it used to go fast and you'd get production out and you felt you did a day's job when you got home." For Lois, this sense of accomplishment was enhanced by her position as "repair girl." Of the 26 women on her conveyor in the late 1940s, Lois was picked by her line supervisor to learn all jobs. Explaining this practice, she stated that "the line would never stop. If anybody had to go to the restroom or anything, we'd take their place." The job of repair girl was particularly beneficial in that it

Figure 2.7. Assembling timers on the conveyor at Sessions, circa 1950. SESSIONS, *127 YEARS OF CLOCK CRAFTSMANSHIP.*

Figure 2.8. Automatic Room at Sessions, circa 1950. SESSIONS, *127 YEARS OF CLOCK CRAFTSMANSHIP.*

afforded protection against layoffs. "I would not get laid off because I knew all the jobs," she explained, boasting "I was a '10 percent' of the help."[2.33]

While favoring conveyor work above other tasks, Lois also liked packing, which she believed offered a similar feeling of satisfaction in doing a "day's job." The gratification that she got at the end of the day on the conveyor, she reasoned, "was the same thing with packing." In contrast, Lois liked press operation the least because it did not transmit the feeling of accomplishment as the other positions:

> When you work on a machine, you've got so many to do and it stayed there, [so] you don't know if you did. At that time, you didn't know what you put in there. You just kept pushing, and there was times where the machine would break down and you'd get upset that you didn't know if you made production or not.

Agnes Moquin, who worked on both the conveyor and presswork, echoed Virginia's disdain for presses. Agnes focused on the stationary aspect of the job and described work on pedal-operated presses for springs as physically restrictive. In particular, she talked of how, to make this type of press function, a woman had to be "chained" into it. Such a bias against presswork at Sessions differed from Ingraham, where women did not single out this labor as unpleasant as did Agnes and Lois—perhaps because of the absence of the comparison with conveyors there.[2.34]

Whether at Ingraham or Sessions, working women learned new identities that grew out of their single-sex job environment—ones that challenged the image of being just "girls," as they liked to call themselves. In a work group that is exclusively female, wage earners forge bonds of solidarity, mutual respect, and understanding of their skills and competence as workers. Solidarity, mutual respect, and common identity took numerous forms. Employment in Bristol's highest-paying factory in the 1930s instilled a sense of success, which the women equated with a continuing education after high school. As Kay put it,

> We [women] used to call it 'Ingraham College,' to tell you the truth, in my days, because all the girls would get out of high school, some would go to work in the office, some did not. They'd go directly to the factory. They made more money working there because in the office they didn't make as much money.

For Kay, the "college" aspect of work also included the contacts made in a single-sex environment. "To me going to Ingraham was like going to college. It was a girls' factory and it was just like a girls' college now. They run in and out and talk to everyone. It was the same thing for me. That was my education."[2.35]

This type of common identity allowed coworkers to expand their experiences outside of the family when not on the job, and in some instances beyond Bristol, where their bonds of kinship lay. When asked how she felt working in the all-women balance wheel department, Kay gushed:

> That felt great because I joined the bowling league and we'd go out for dinner once in a while and we'd go to see a play. In those days you could get an excursion train to New York for two dollars and then we'd put a dollar a week away and when we got ten dollars. . . we'd go see a play. I did that every so often and I was the one who would collect the money from the different girls.

Kay used her employment as a source of empowerment, deriving a stronger sense of self from her afterwork pursuits. Women office workers from various departments throughout the factory behaved similarly when they formed the Ingraham Business Girls Club in the early 1940s to enhance their understanding of the traditionally male realm of business practices.[2.36]

Cementing the bonds of solidarity among working "girls" was the appropriation of exclusively female space

Figure 2.9. Sessions women working on electric clock motors, Christmastime, circa 1960.

and time on the job and off. Both men and women developed gender-specific spaces that merged aspects of their workdays with their off-the-job lives. To break the daily monotony of factory production and gain more control over their work, men and women talked to each other and participated in customs that identified them by their gender. For example, men collectively carved out male space for themselves at work through the common male practice of chewing tobacco. "All the fellows—90 percent of them chewed tobacco—had their spittoons," explained John Denehy to describe watch adjusters. "Tetro [the head foreman and Bob's father] chewed. They all chewed." Men also displayed healthy appetites for playing practical jokes. Among the most common of such antics was nailing other men's work shoes to the floor. Jerry Nocera related the typical scenario:

> They had [their shoes] sitting in the corner when they took them off, put their clean shoes on after the day's work was done, and they'd take them off and then when they're gone, we just put a spike through the bottom. He'd never know. They get there in the morning, shove their feet in there and when they start walking, they didn't notice the nails in there.

Jerry explained the motivation behind such tomfoolery and in the process illustrated another popular trick among men as well:

> Question: What was it done for?
> Answer: Something to reminisce years later, you know, and also, when they were in the men's room while they were sitting on the bowl or something like that, you'd throw water over the door. Something like that, yeah. Oh, we had—yeah, we had a lot of fun after all. That's good natural fun. What the hell.[2.37]

Men from both firms remembered talking at work mostly of sports. Reinforcing this sports culture, males frequently joined company or departmental teams that played after work and on the weekends and served as additional topics of workplace discussion. For men eager to prove their self-worth, participation in sports was a must. At Ingraham, teams were organized by department, allowing workers to challenge the gradations by job and pay and compensate by maybe beating a higher-paying department. "Department against department and so forth," John recalled. "That was one way you fought each other [laughs]." At Sessions, teams were drawn from men throughout the company. Jerry played on the "softball" team and remembered that the men came from all over the firm. "They come from the toolroom. They come from the spray room. They come from the—even had them from the office." These men tested their self-worth on the playing field not against each other but men from other local teams, against which Ingraham's teams also played. No special rivalry, however, existed between the two clockmakers.[2.38]

Women also created their own cultures, which included a share of pranks—although these were less physical than the men's. Mary Penoncello, who worked at Sessions for six months in 1937 and 1938, recounted how she used to tease another woman, who like herself racked clocks to put them on the run. "I used to drive one lady crazy. She was kind of a big lady [laughs] and everybody wore aprons because you can get dirty, you know, and I'd go behind her and untie her apron because I knew she couldn't reach it to tie it up again [and she'd say] 'I'll get you for that.'" Similarly, women, like men, took time to talk at work about gender-specific topics. Kay Laviero remembered the typical conversation of women in the watch department:

> Question: What did you talk about?
> Answer: What did we talk about? Boys, clothes, going out. 'Did you buy a new outfit today? How did you spend your money?' Things like that. Putting make-up on, you know, putting lipstick on. . . . that's what we did, talked about clothes and the boys . . .

Such talking dominated the conveyor belt at Sessions. "All we did was laugh and talk," related Kitty Baldaccini. "And we'd play games while we were working," she went on to say. "We'd play games about, we'd give initials maybe of an actor or an actress and we're supposed to guess who it is. It was fun, it was really nice."[2.39]

Some did all they could to infuse their job with a gendered spirit. By the 1930s the mostly women service department at Ingraham was under the direction of a less than stern male head. Former repairer Patricia "Pat" Letizia related how two women in particular took advantage of the situation:

> They'd let their work pile up on their desk and everything. Two girls worked . . . right near the boss and they made a whole afghan. . . . They'd take turns going to the ladies room—and they made it. The two of them worked on it [there] and when they made it, they raffled it off. They even had the nerve to ask him if he wanted to buy chances on it. [Also], they'd go out

and... wash their hair.... If they had a date, they'd go out and by the side of the ladies room there was a sink and they'd bring their shampoo and they'd wash their hair, and then... another girl would go out and she would put their hair for them in bobby pins.

Pat was shocked that the boss never confronted the women. "He must have seen them... but he never said a word to them."[2.40]

Work Cultures in Conflict

Despite their distinctive single-sex natures, women's and men's cultures occasionally met head on at work and reinforced separate gender identities. While an inspector on 30-hour alarms at Ingraham during the 1930s, Bertha D. had four women under her in a department that neighbored the all-male spray room. If the women left their work shoes out at the end of the day, the men sneaked in, painted the shoes, and when the "girls" went to use their footwear the next morning, they'd find them red, blue, or yellow. Bertha recounted that the women found the joke mildly amusing. Because they targeted the women collectively, the pranks likely also fostered solidarity among the women against the male sprayers, as Bertha implied was in fact the case. For men, too, contact with the opposite sex at work reinforced their own identities. Men in Ingraham's toolroom, for example, regularly whistled from their windows to groups of women coming and leaving work at the factory. This whistling became an integral part of their work culture and thus of their identity as men. Enhancing the male value of this action was the threat of reprimand from a foreman if reported by the women or woman targeted, which apparently wasn't an uncommon response.[2.41]

Truckers in all-women departments especially tried to reinforce their maleness against women because, operating singly, they were most vulnerable, or most directly exposed, to large numbers of women. Before taking up toolmaking, Carl Kirkby worked briefly as a trucker for assemblers on 30-hour clocks. "I had to keep six or seven girls, maybe eight, supplied with parts as they were assembling clocks," he remembered. The women threatened Carl because they both outnumbered him and did not accord him the docility and respect he wanted. Coming "from a home where you had to be a gentleman and have manners and act halfway human," Carl explained that "going to work with those eight women was quite an experience." "They had me running," he continued, "because they were on piecework and they had to make things fast in order to make good money." To counter the affront that their aggressive behavior posed, Carl decided to exert what he believed to be manly persuasion, and in the process acted also like a father training children to behave properly—not unlike how the owners approached their workers:

So I tolerated that for maybe one month and I thought to myself, 'Boy, I got to get these women straightened away.' So I said, 'Girls, from now on the ones that say "please" get served first.' 'Like hell. Like hell. We don't say please for nobody.' I said, 'Here we go.' ... They finally realized they had to say please if they wanted anything. So I encouraged them to say thank-you afterwards.

Despite his apparent victory, Carl conceded that he was not completely successful; one woman continued to "irk" him, so he retaliated by giving her a present at Christmas of cockroaches, which he knew she loathed.[2.42]

The experiences of men and women also differed in that women did not invoke hierarchies at work. Working women were aware of the competition that characterized men's work, especially when exposed to males who performed the more-skilled labor. In the watch department, for example, assemblers sat beside highly paid adjusters, so engrossed with their labor that they seldom joined in even the talk that punctuated factory life. The women did not challenge the men's silence. "There was a girl that sat next to me, [who] talked constantly all day long ... ," recounted Phil Carrier of her work on pocket watches. "And I said, 'You know, Lil, you shouldn't be talking. You're distracting the adjusters in the front. And these guys can't take it.'"[2.43] However, while acknowledging men's hierarchical structures, women did not follow suit. Instead, they strengthened the already nonhierarchical character of their unskilled jobs, emphasizing their common out-of-work identity as family and community members as the source of their womanhood.

"Helping," for example, was a custom that was common to women at both factories but unheard-of among men. Growing out of the nonhierarchical webs of women's work at home and in the community, this defining practice of aiding one another in getting out production was prevalent in most all-women departments. Interviewees from Ingraham talked of fast workers helping slower individuals in the service, watch, and 8-day assembly departments. Even on Sessions' conveyors where each woman had a different task, helping was common. Lois Cieszynski noted that "we all tried to help each other and go right down the line to meet production."[2.44]

In at least one department at Ingraham, women extended the notion of helping to the stint, the informal worker-set production quotas that apparently survived until the demise of the industry but only among women. Walantyna Sakowski related that she and her 1940s co-workers in 8-day assembly "used to set their own quotas. ... We didn't try to compete. We did what we could." In such an environment, which the company's paternalistic management sanctioned, talking and singing were tools of helping. Sabrina compensated slower workers by trying to help out "if I'd see that they were struggling." However, she also recalled one worker who "was slower than I and we used to sit there and we'd sing and talk." Walantyna saw this combination as helpful, stating that for both women, "the day went by so fast" as a result of their interaction. In her particularly egalitarian department, Walantyna could remember little antagonism among coworkers and reported as one example another woman's skepticism that she was a true blonde (which she was). Evidently, competition limited itself to off-the-job aspects of feminine rivalry.[2.45]

The Effects of World War II

The exception to helping and other norms of workplace relations was World War II. The mandates of wartime production hindered or altered those relationships at least temporarily. At the close-knit Ingraham, for example, the presence of unfamiliar faces of outside inspectors from the government interrupted everyday life. "Then every so often the [fuse] inspectors from Washington, DC, would come to our room and they'd be on your back," related Kay who added, "I used to be a nervous wreck." She explained the source of her nervousness when asked what the inspectors wanted to know. "Well, he wanted to see how I was working and what I was doing and he'd ask me questions. Oh, my God, I thought I'd die from fright." She was able to answer questions well enough:

> They would say, 'Are you finding any rejects? Are there any burs on it?' I'd say, 'There are not too many,' and the ones that were bad, I had to put them to one side and he would say, 'How many bad ones do you have throughout the day?' I'd say, 'Well, not too many,' but maybe in a hundred you'd have one.

Under such tight control, the ability to help one another get out production would be stunted.[2.46]

Likewise, defense work created a fear of nonconformity that the Cold War era later magnified. Given a high priority rating by the government, Ingraham instituted strict rules for entering and leaving the plant in May 1942 and likewise issued workers badges with photos and numbers, checked on entry. By government order, Ingraham also identified its non-citizen workers—33 individuals by early 1942—and later barred all "aliens" from transfer to any defense work labeled "Classified Contract," despite the fact that the fuse program had been pioneered by the toolroom's Ferdinand Fengler, a German American. Although Ingraham's union defiantly filed a plantwide, but unsuccessful, grievance against the "alien angle," the wartime climate bred fear and pressures for conformity. Al Calderoni, who routed fuse plates during the war on his climb to the foremanship of the electric department, talked uneasily about the consequences of nonconformity, mentioning "one fella that they actually took out of there." "He worked in the service department," said Al, "and someone came in and walked right out with him. Never heard of the fella any more." Al said nothing more on the subject, not even guessing what prompted the employee's removal. His tight-lipped stance about what happened years after the occurrence suggests the persistent cloud of fear that hung over factories like Ingraham, heightened by later memories of McCarthyism.[2.47]

The war also brought a more permanent change. It allowed returning veterans to shape new perceptions of themselves vis-à-vis other workers that resulted from the soldiers' unique experiences abroad and that in turn altered traditional workplace relationships. Carl Kirkby, who served as a combat engineer with the First, Third, and Fifth Armies in Europe, participated in the liberation of the Buchenwald concentration camp. The horrors he confronted undermined the time-honored hierarchy of age and skill that dominated the toolroom after his return. Only about 26 years old, the young veteran challenged tradition by shaming an older worker, Ed Coughlin, who was "older than even my dad" and who questioned the reality of the Holocaust. Carl liked the "Irishman" well enough, citing, "We got along great." He also pointed out that, as a senior employee in the room, Coughlin had the privilege of eating at his bench in any manner he saw fit: "Every day at noontime he used to spread a little handkerchief, a little napkin on his bench and he'd open up his thermos of coffee . . . and he'd open up his little sandwich and sat there eating it."[2.48]

Still Carl felt compelled to challenge his elder one day. "So we were talking about the war, a few of the fellows and I, and I was telling them what happened at Buchenwald," Carl recalled. "What I

Figure 2.10. Kay Laviero's badge.

Figure 2.11. Dottie Beaucar's World War II security badge, Ingraham, 1942.

seen. Took pictures of them [the bodies in the camp]. So [one of my co-workers] says, 'Go in and see Coughlin. . . . Coughlin insists there never was a Buchenwald.' Carl, who years later in his interview cried when talking of his experience at the camp, was shocked and determined to set the older man straight. He brought his photos the next day and waited for the right moment:

> Yeah, well, anyway [Coughlin] was just opening his own act up there and all that sort of stuff. So I waited until he got his sandwich in his hand ready to bite, and I put these five pictures in front of him of the bodies piled up in groups around the concentration camp and the furnaces where they burned them alive and all that. He says, 'Where did you get these from, Carl?' I said, 'I took them. So don't try to tell me it never happened.' He says, 'Well, if you hadn't showed me those, I'd call you a damn liar, but I know you're not a liar. You must have taken them.'[2.49]

Carl's hostility reflected the right of rebuke that he felt his firsthand experience earned him, but it also revealed a change in authority patterns as a result of the war: the young man behaved in a manner that previously would have carried serious consequences (as Carl knew personally from his experience with Mr. Osgood). His actions suggested that wartime experiences eroded the advantage that age and skill held over younger men like Carl who had been trained as apprentices to accommodate and ultimately reproduce this hierarchy.

Conclusion

The war years altered the workplace and temporarily curtailed clock and watch manufacturing. That period aside, however, daily life focused on the production of clocks and watches. Setting the stage for fuse movements, the work entailed the manufacture of numerous pieces, all of which needed to be finely finished prior to final assembly. Both Ingraham and Sessions made these parts—from screws, wheels, pinions, and plates to cases made either of metal or wood. To manufacture the cases and movements in turn required the fabrication and/or preparation of other materials: complex tools and dies for movement parts, and in the instance of wooden cases, sandpaper, sprays, and other finishing materials. All these jobs fell to men and paid on the basis of skill. In a family-like atmosphere, workers passed their skills to successors

Figure 2.12. Sessions' float in victory parade at the end of World War II, 1945. Behind is the factory's main office. Laura (Mastrianni) Santago, top right. Her sister-in-law Aura Mastrianni, then Nocera, top left.

they deemed worthy through patronage and deference. In this way, social hierarchies replicated themselves from generation to generation. At Ingraham, this complex arrangement extended into the realm of watch adjusting, a post-assembly job that paid among the highest wages in the company.

Although men dominated woodworking and the production of the necessary tools for clocks and watch manufacture and in the process carved out multilayered niches for themselves, women performed the presswork, assembly, and most finishing procedures on timepieces. These jobs were far less hierarchical than men's. Women floor workers did develop hierarchies of their own, judging one position better than another. However, they worked less competitively than men, often helping each other meet quotas and sometimes setting the quotas themselves, strengthening the nonhierarchical relationships in which such actions originated. All the while, they saw themselves as "girls" and not women. Unlike men who derived their identities from the wages and skills of their craft, women saw work as only one aspect of their lives, despite the long service records that individual employees would sometimes amass. Seeing their commonalities off the job led to far less stratified environments at work. At the same time, it allowed women to become more than just docile "girls," even if they never stopped referring to themselves as such.

PART II
Change and Its Meanings

Chapter 3
Workers Take a Stand: CIO Unions at Ingraham and Sessions

In 1941 the Congress of Industrial Organizations (CIO)'s United Electrical, Radio, and Machine Workers of America (UE) successfully completed a three-year campaign to organize Bristol's clock and watch industry. Sessions and Ingraham followed other Bristol firms manufacturing brass, locks, hardware, and ball bearings into the CIO fold. The paths of Ingraham's and Sessions' locals, however, diverged quickly. Sessions killed off the union in 1942 during the conversion process to wartime production by laying off all members of Local 264. Ingraham's Local 260 consolidated its victory but during the second half of the 1940s also faced challenges. The Cold War had a chilling effect on the union. Postwar paranoia over communism infested the firm and targeted its union as un-American. Hyperpatriotic fervor against the perceived menace of an international communist threat precipitated a bitter rivalry between UE Local 260 and an internal right-wing faction that wanted the International Union of Electrical Workers (IUE), which had replaced UE in the CIO in 1949. The outcome of the struggle left visible scars in the working-class community and brought a new type of unionism to the factory, one that took a more accommodating tone with management than did the UE's "them and us" version.[3.1]

The CIO in Bristol

The CIO first appeared in Bristol in July 1937, with the protection of the Wagner National Industrial Recovery Act of 1935, which lifted antitrust regulations and endorsed trade associations. Even so, the CIO and industrial unionism did not find an easy entry into Bristol, owing to the usual associations of unions with radicalism, and in particular with the Communist Party. The last attempt to unionize workplaces on a citywide basis had happened during World War I. Then, two Italian immigrants and two Russians representing the radical Industrial Workers of the World (IWW) tried to organize Bristol factories under the "Wobbly" agenda of a dialectical fight to the death with capitalism. In the heated climate of war, a combination of xenophobia, patriotism, and invocation of sedition law led to the arrests and swift convictions of all four. Despite a less radical outlook, the CIO fared only slightly better 20 years later. From the outset, the CIO focused its organizing efforts upon the city's two largest employers, Ingraham and the massive New Departure Division of General Motors. However, local halls refused to allow activists at either company to hold public meetings on behalf of the CIO.[3.2]

The CIO also faced apathy from workers themselves in their respective efforts at New Departure and Ingraham. By December 1938 New Departure employees had not yet been able to get even the 200 signatures necessary to secure a charter for a local with the United Auto Workers (UAW).[3.3] At Ingraham, such apathy can be traced to the strong worker identification with the firm. Steady work, increasing wages throughout the Depression, and the recent introduction of paid vacations and Christmas bonuses for all employees almost certainly undermined initial action on behalf of the UE.

By 1940, however, the CIO in Bristol had successfully stirred worker interest. At New Departure, workers eligible for union membership gave its UAW Local 260 a 20-to-1 victory in the June election held by the National Labor Relations Board (NLRB), the governing arm of the Wagner Act. The UE made gains as well, not only getting a foothold in Ingraham but also establishing a local at Sessions by January 1941. In part, the UE success can be traced to the appearance of an organizer gifted at presenting himself as working for common cause with the local community. This individual, Joe Caiazza, stayed with the union campaigns at both Ingraham and Sessions until they were brought to a successful close.[3.4]

An Italian-American from nearby New Britain, Joseph Louis Caiazza was 32 in late June 1940 when he applied for, and was accepted as, a field organizer for the UE. For two years previously, he had worked as an inspector and assembler in the electric iron and heater departments at New Britain's Landers, Frary, and Clark and had served for eight months as a business agent for both Landers, Frary's UE Local 245 and another local also in New Britain. An active organizer from June 1940 through July 1942, Caiazza spearheaded organizing tactics not only in factories in Bristol but throughout Connecticut, including General Electric plants in Meriden and Trumbull; Hartford's Colt Firearms, Gray Manufacturing, and Underwood Typewriter; Bristol's Union Hardware; and a factory in nearby Thomaston. Caiazza honed his organizing skills by taking a labor course in New Britain. Articulate and well versed in unionism, Caiazza manifested a strong commitment to CIO activism at Ingraham, where he already had contacts in June 1938. Charlie Rivers, the UE's field representative in the Bristol area and a supporter of Caiazza in his bid for organizer, reported that he spent the majority of the week prior to applying for that position with a group of workers from Ingraham. Troubleshooting the problems they presented, Caiazza apparently interacted well with the group and won its backing.[3.5]

Beyond the emergence of an able organizer, the UE also benefited from the subtle feeling among many workers that unionization was inevitable. John Denehy, the watch adjuster and later superintendent, recalled the attitude. Thinking back to the start of the 1940s, John reasoned that "unions were coming:"

> Unions were getting stronger. Organization, that was the coming thing. So before you got through, almost every factory in town was organized. Then it became where your politicians or your city help became unionized. . . . . . It went all the way up the line. People . . . that you never would think would be organized.

John's words should not be taken to mean that workers were simply swept up by "organization," or what UE Archivist David Rosenberg has called the "CIO steamroller" of 1940 and 1941; employees made choices on their own and often opposed unionism, particularly in the case of Ingraham. Nonetheless, most workers knew of the CIO victories achieved in Bristol and throughout the nation, either by reading the *Bristol Press* and/or through word of mouth. A resulting belief of "why not us, too" cannot be ruled out in motivating worker activism on behalf of the UE.[3.6]

UE Local 264, Sessions Clock Company

The steamroller effect drew the attention of Sessions workers to the idea of unionism in 1940. Mary Budnik, Kay Laviero's sister and a timekeeper for women press workers at the company in 1939 and 1940, didn't even remember UE activity when she worked there. "I never heard any rumors or anything or any inclinations of a union trying to get in," Mary exclaimed because "everybody at Sessions was satisfied."[3.7] However, Sessions' employees, especially those who worked there throughout the 1930s, were in fact more open to unionism than Mary's words indicate. The unstable condition of the company placed workers in a tenuous economic position, with regular short time and the accompanying decline in their incomes. As a result, the UE inspired immediate interest once organizing began, especially in the unskilled departments of the case shop where wages were lowest. Working from age 16 in 1934 until laid off in 1942, Jerry Nocera was among Sessions' first union activists and served as the first steward of the spray room. "Well, I was in the union too, yeah," Jerry proclaimed. "I was a steward of my department." Jerry's initial account of the beginning and success of the UE at Sessions was vague. "I had very little recollection about that [the origins of the union at Sessions]," he said:

> It started, I guess, by people from the outside, they wanted to start the union and they got a hold of a few of the employees and they got them to pass the word around: Who wants to join? See, if the majority of them would like a union and apparently they did, and once they filed for the union to get in, they presented their papers and they says, 'Well, the majority of your employees want to belong to the United Electrical Workers.' So we had an election and we got in.

When questioned further, however, Jerry clarified where he believed unionism started in the company. "I would say it started in the spray room." When asked why, he responded, "I don't know. I guess they were mostly all men, as far as that goes, and they're the ones that I guess were the ones that started everything."[3.8]

Jerry's words here hint at the origins of the Sessions activism. Men, however, also had the support of their working sisters in this family-oriented factory. Promoting a cross-gender sense of community, the plant's close-knit structure made joining a union a family act. Once open to unions as "the coming thing," workers were receptive to the UE, even though it was an outside organization. This attitude stemmed primarily from acquiescence on the part of the manager/owners once organizing began. The firm's silence was peculiar but can be explained in that the company's history of economic weakness and its receipt of the Reconstruction Finance Corporation loan undercut managerial ability to counter organizing efforts. The company's debt to the federal government, which now endorsed unions, in particular bound the Sessionses to compliance with legislation like the Wagner Act. However, W. K. and "Bud" Sessions, the father and son team who ran the company, were not pleased about an outside organization representing their workers. Frank Savage, the former production superintendent, was close to Bud Sessions and touched upon the younger Sessions' discontent when discussing the NLRB election at the company. "Well, young Buddy didn't like the idea of converting to a union," Frank recalled. "He figured that the Sessions family had had the name for years [i.e., had control of the company] and they wanted to keep it that way, but when they finally took a vote on it . . . they voted to have a union."[3.9]

The outward silence of the Sessionses impressed activists like Jerry who reported "not too much" resistance from management, and boasted, "We never had confrontations with anybody, regardless of whether it was inside or outside, you know." Consequently, Jerry thought the company supported unionization: "I think they went along with it. At least they didn't show it." With the supposed approval of management, animosities between those who supported the UE and non-unionists were few. Frank, who was not eligible for union membership, portrayed the UE campaign as tranquil, with individual choices respected. "Oh, they was trying to convert everybody over to a union," Frank said, "and some people would go and, well, like myself, I didn't go. I didn't belong to the union." Similarly, according to Frank, union activities were open. Because workers believed the owners supported them, unionists forged their local as an extension of the company family.[3.10]

For this reason, even the most ardent unionists bided their time in building Sessions' UE Local 264 as a class-based, as opposed to a family, organization. "So every so often," Jerry explained, "we [stewards]'d hand in more cards to the union headquarters and we gradually got up to where we could demand more for our labor, see?" Jerry was proud of the local's record, even if the union chose not to provoke management. As he concluded,

> we did all right, though. We did all right. Our wages went up and we got our vacations, which we didn't get before, and I don't recall getting any insurance. That I don't recall if they gave us insurance or not. Perhaps it was because of the fact that the union just got in and they weren't too demanding, you know, what I mean, until they got stronger, you know what I mean?

Because of the noncombative stand taken, even the pro-boss *Bristol Press* recorded an amiable relationship between the owners and union upon the completion of UE's first contract in April 1941. Represented by its president Albert Westfall and organizer Joe Caiazza, the local reached an agreement on contract provisions with the company after six weeks of closed negotiations. In a public show of support for the contract, company president W. K. Sessions issued a joint statement with Caiazza and Westfall that declared the contract "should make future relations more harmonious and better for both parties." With this optimistic outlook, the trio ushered in the formal operation of the short-lived UE Local 264 at Sessions Clock.[3.11]

Ingraham and UE Local 260

Two weeks before the announcement of the first union contract at Sessions, Ingraham's own union, UE Local 260, won the right to bargain collectively for employees at the Bristol factory with 52.5 percent of the votes cast. The results of the NLRB election on March 26 marked the first time in the company's 110-year history that an outside organization mediated workplace relations. Emboldened by Ingraham's huge growth during the 1930s, the owners hoped to withstand the union fever sweeping the city. As a sign of strength, they had even begun expansion beyond Bristol for the first time in the firm's history. On March 13, 1941, the company announced that it had acquired the Thiel-Canadian Watch Company of Toronto, reorganizing it as the Ingraham-Canadian Clock Company, Ltd., to serve as a watch and clock assembly unit for the Bristol-based firm. The union's success in the wake of this bold move represented a very unwelcome departure for the Ingrahams.[3.12]

Referring to UE organizers as "inexperienced outside men from outside the industry," the owners openly opposed unionization throughout the campaign at their factory, if only through the written word. Before the election, the Ingrahams addressed a letter to the workforce that made their position clear: "Any representation that the Company favors a Union is not the truth." Citing the full gamut of employment policies, the brothers argued their paternalist style to benefit both workers and management: "The Company has been happy in its relations with its employees, with the personal contacts it has enjoyed with them, and it does not desire a change from this close relationship. We feel that we have both worked together, shoulder to shoulder, and have gained many advantages for both of us in this personal relationship."[3.13]

However, by March 1941 when the Ingrahams issued this letter, a particular group of men did not feel the same way. These men, who labored in the company's primarily male case shop, gave the UE the support it so wanted. "It was the case shop that really started the union," John Denehy explained. "Al DeCapua [a leading unionist and from June 1942 onward the longest-serving president of the local] was the number one guy down there and they weren't paid that much down in the case shop." Bob Tetro, the former executive vice president, concluded similarly, citing, "The case shop was easy for them to organize because everybody down there was mad." Edward Ingraham himself, in his personal memoirs, expressed the belief that the union started in the cabinetmaking end of the factory. Ingraham wrote, "In 1941 our employees were unionized—the Union starting largely through dissatisfaction among the employees in our Case Shop."[3.14]

The "dissatisfaction" solidified through a convergence of several factors: low piece rates, unsteady work—despite a booming business in radio cabinets for Emerson and Firestone that took up 25 percent of plant space by 1937—and ethnic notions of maleness. At the center of case shop employment was a group of Italian American men who comprised approximately 44 percent of the division's 298 men and women. A high proportion labored in the cabinet and sanding department where most of the actual woodworking occurred, accounting for 64 percent of the total employees. So dominant was the ethnic presence in the case shop that Sadie Witik, the division's native-born timekeeper, had to learn Italian to communicate with most of the workers.[3.15]

Trained in a family tradition of highly skilled woodworking, Italian American men in the case shop equated their masculinity with the ability to make finely crafted woodwork. They saw themselves as artists whose individual competence was essential to their self-worth. Oreste DePascale, the former foreman of the cabinet and sanding department, put this value system into words. Talking of his ability to do woodwork, which he learned as a boy in his native Bagnoli Irpino in Italy's Avellino province, Oreste boasted, "Whatever I did over here [in America] was my idea. I never copied anybody." Work at Ingraham contradicted the equivalence of masculinity with expert woodworking skill. On the job, the division of labor dictated that workers produce copies of patterns provided to them, leaving little room for expression of individual artistic abilities. Additionally, shop employees increasingly performed their work on machines, not with their hands, whose skill they prized.[3.16]

Shop workers, of course, met job deskilling by redefining notions of masculine identity in part as the ability to meet the production norm. However, piece rates complicated this goal by denying them the artistic ability to take the time they felt necessary to execute effectively their woodwork. Cabinetmakers and sanders voiced frustration over low piece rates in a letter of March 17, 1940, to Edward Ingraham in which they announced their intention to join the CIO unless conditions improved. In this first formal communiqué from Ingraham's workers on behalf of unionism, the employees wrote of "the low wages yours [sic] Company is paying us." "We work peace [sic] work," the letter read, "often less than for 40 cents an hour and we have no time to go to the toilet. . . . This is the reason most of us want to join the C.I.O. and the union will probably protect us." Signed the "Employees of the case shop," the letter ended by implying that unionization could be avoided if piece rates were raised to an acceptable level so that workers were not so rushed. Despite some adjustment, however, most piece rates remained unchanged.[3.17]

Lack of work and frequent layoffs compounded the effects of low wages and fast-paced production. Tommy DiSabato, a machine operator who was trained in Italy as a gunsmith and found woodworking easy in comparison, left the Winchester Repeating Arms Company in New Haven in 1930 to join his father at Ingraham. Tommy's work schedule, however, consistently followed a pattern of a short time of only two or three days a week throughout the 1930s. "It was better to stay over there [at Winchester]," he exclaimed. "I should never [have] come here." So persistent was the trend of unsteady work for Tommy that in 1938 he went back to Italy for a few months. "I go there on vacation and then I get married," he recalled "Because no work over here. What I say, I get two days a week. Two days. Three days. Because then there was a slack over here. . . . I stayed six months in Italy. And I work in Italy. I find my wife and I get married." Tommy returned with his bride to Bristol and to Ingraham where he remained for the next 30 years, laying down roots and getting steadier work.[3.18]

Most other Italian Americans shared Tommy's status as a married man. For them, the lack of work and low wages was particularly emasculating. From Italy, they had brought the idea that a married couple worked in a partnership of sorts, with the husband earning money while his wife did unpaid labor inside the home. Oreste, who married an Italian immigrant from New Haven, summed up this notion. When asked if his wife worked, he responded:

> When I got married I don't want the woman to work. I didn't want my wife to work because I don't believe in woman's work. A woman has got plenty, plenty of work in the house. Cleaning the house. Wash. Cook. After the kids come, she got plenty of work in the house. I never believe in the woman's work. That's wrong. That's wrong. The woman has to stay home.

Oreste went on to say, "We, in the old country where I come from, used to say, 'The woman inside and the men outside.' In other words, the men got to earn what the family needs outside, you know, and the woman inside." Because he brought home a foreman's pay, Oreste successfully kept his wife at home. However, Italian American men beneath him were often unable to provide a living wage for the family and lost the masculine role of sole wage earner. Having wives work, and perhaps children as well, was enough of an affront to motivate interest in the CIO. The union came, Tommy said, because "you know, the job, they [Ingraham] didn't want to pay. Because a lot of people, there was all married men. They got kids."[3.19]

Exacerbating problems with the existing wage structure and cementing union support were the division bosses themselves, and in particular the shop superintendent, Ostilio Ciccarelli. Ingraham brought Ostilio Ciccarelli to Bristol in 1928 as foreman of the sanding department, having lured him from the Victor Talking Machine Company in Camden, NJ, where he had worked for 17 years. Responsible for designing many of Ingraham clock cases and later Emerson radio cabinets, Ciccarelli, or "Chic" as he was known throughout the plant, earned the respect of company officials and was quickly promoted to shop superintendent. Trained as a cabinetmaker in Italy, Ciccarelli, in the words of Bob Tetro, "knew very well the intricacies of bending wood. We did a lot of bent work, use of veneers of finishing, and from that end of the business Chic was very knowledgable." "In terms of handling people," Bob added, "Chic was almost impossible."[3.20]

Described by Edward Ingraham as a "'little Mussolini' who had little feel for labor relations," Ciccarelli operated his shop by constantly driving his help through such tactics as hollering, threats, and swearing, almost always communicating with his employees by yelling at them. "You could hear him on the street," remembered Tommy. "And I had a tough life with him," Tommy complained. "Ciccarelli there get [sic] what he wanted. Ciccarelli was

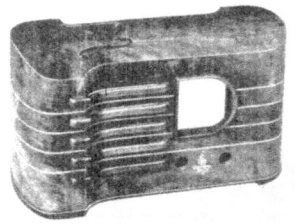

Figure 3.1. Emerson radio and television cabinets, characteristic of Ciccarelli's bent wood designs, from the 1939-1940 Ingraham catalog.

like the owner. And when [he] say something, you better do." Workers tried to avoid conflict if possible: "But maybe if you don't answer back, maybe he won't holler," as Tommy put it. He knew too well the consequence of responding to an order from Ciccarelli. "I answered," he explained: "You know, he come over there [and said to me], 'Make the radio.' He come after two or three hours, 'Who told you to do this?' I had to say, 'You told me to do this.'" Ciccarelli then began hollering at him. In another instance from March 1940, Anthony DiLorenzo, a rubber who with Albert DeCapua was the chief proponent of the CIO in the case shop, argued with Ciccarelli about not being able to make out at the set rate and thus work for the price. Ciccarelli summarily dismissed DiLorenzo, cursing him as "no God-damn good, a trouble maker, and a strike starter." The superintendent even took out domestic frustrations on his workers, or so Tommy maintained. In an effort to punctuate the meanness of his boss, Tommy explained, ". . . Ciccarelli must have had trouble with home. When he come [the] last time [it] was with a scratch, he got a fight with his wife and then he pick on us." "He was mean," verified Kay Laviero, recalling her teenage years in the late 1920s as one of the shop's few women. "Whenever he saw one of the men talking to me, he'd send them away. He was so mean I did not like him."[3.21]

Ciccarelli was equally hard on his foremen, pushing them to drive their workers to make money off his designs. Because of such constant prodding, at least one foreman, Oreste DePascale, was often not even on speaking terms with Ciccarelli, either on the job or off. In an atmosphere described as "You do it because I told you," Ciccarelli abused his foremen who in turn pushed their workers to produce. "As a labor relations man," Edward Ingraham later lamented, "he was a complete loss."[3.22]

Ciccarelli's style of management demolished the unstable case shop foundations of Italian American masculinity because of the potential for personal animosity between the bosses. A dismissal from the superintendent himself was particularly insulting. Ciccarelli's talent lay in his ability to design popular cabinets. As a consequence, he was seldom involved in the actual process of woodworking in the shop. Because his subordinates did not see him at work, both foremen like Oreste and workers like Tommy believed Ciccarelli knew little about wood and had no right to judge others. So strong was this sentiment that when Oreste, in the role of foreman, offered a highly regarded finisher named Frank Capone the foremanship of the finishing department, the worker declined on the basis that Ciccarelli wasn't worthy to be his boss and instead went back to his native Italy. Oreste remembered Capone saying, "'Under Ciccarelli, I will never do.' He says, 'Never.' He says to me, 'I don't mind when you got the foreman who knows more than you.' He says, 'Ciccarelli don't know anything about it. Why should I take all the things he say[?]'" Such an attitude was reinforced in the spring of 1940 when a trucker named Frank Caracciolo was advanced with no hands-on experience with wood to an inspector in the finishing department, above experienced rubbers and finishers. Resentful, the men protested that they were more qualified than Caracciolo, "who didn't know rubbing or finishing," and became more determined to organize.[3.23]

Left to his own style of management by the paternalist Ingrahams, Ciccarelli continued his antagonism leading up to the election, further committing case shop workers to the UE. Once pro-UE sentiments jelled in the shop, Ciccarelli became openly hostile toward union members like Anthony DiLorenzo, in violation of NLRB regulations and counter to the paternalism the Ingrahams sought to enforce. Tommy reconstructed Ciccarelli's behavior when asked who spearheaded union efforts. "DiLorenzo," Tony declared. "He bring it [the union] to Ingraham. It was him and DeCapua." Tommy continued:

> . . . And when they brought the union, everybody but Ciccarelli [wanted it]. He had to spy on us. They paid people that don't work. They just spy. They give everything. Who go there. Who talk. If he don't like it, right away he lay off. Because he was the chief. He was, I told you, [like] the owner of the shop.

The spies infuriated the workers, who in response saw themselves not as men but as sacrificial lambs who absolutely needed a union for protection. Tommy recounted that ". . . we was just like a lamb—you can't do nothing.

Figure 3.2. Ostilio Ciccarelli (center), 1955. At left is Edward Ingraham; at right is Dudley Ingraham.

Then after [the union came] you call DiLorenzo or DeCapua. If you got something to say, they fight for you."[3.24] The union was a timely savior for men whose jobs nullified their self-worth.

The Union Outside of the Case Shop

"We always felt if it hadn't been for 'Chic,' we would not have been unionized," Edward Ingraham later penned. "He laid us open to being organized by a gang who were mostly Italian." On one level, Ingraham was correct. Ciccarelli undermined the previously unstable foundations of the Italian American maleness for many of his workers, who already believed their deskilled jobs were inappropriate for people of their class and sex. Driven by a boss for whom they had little respect, these individuals saw unionization as their one alternative to preserve their way of life. Edward Ingraham's argument, however, obscures a wide range of other worker strategies behind unionization. Outside of the casemaking division, the movement shop provided support from men and especially women, without whom the UE could not have secured a plantwide contract. Males, especially the young looking forward to lifelong work, supported the UE in a bid for the breadwinner wages they desired. Women activists rallied for greater job security and family protection, believing strongly in the straight seniority that the UE offered over the paternalistic, male-focused model of the owners. "The union came in," recalled Bob Tetro, who sat on the first negotiating team for management, "on certain basic things that they were going to get, and one of them was seniority, a formal seniority program." Those with husbands at New Departure also took a strong pro-UE stand, encouraged by their UAW-affiliated spouses to help build a strong union base in the city. Accounting for 56 percent of the firm's 1,153 women as of February 1941, or one-fourth of the workforce, married women were proportionally supportive of the union.[3.25]

The union also found support among women, both married and single, who looked for fairness among pieceworkers. For those who valued the solidarity of the stint and those who worked a double day and tried to avoid exhausting overwork at the plant, the union offered the chance to prevent faster pieceworkers from ruining the quotas of the slower—or purposely slower—majorities. For some married women, who had other priorities beyond work, the preservation of on-the-job sociability was particularly important. Lillian Rock, a divorced and then remarried mother working at the company at a later date, found fast pieceworkers offensive. After moving in and out of the workforce to raise her family, Lillian settled into a steady job at the plant after its move to Redstone Hill Road in East Bristol in 1964 and became a steward of her room. When asked why, she explained, "A lot of them [who were fast pieceworkers] would just make it very difficult for a lot of the other girls, and I think that mainly that's one of the reasons why I became a steward in my room." Her experience with the fast women made her a lifelong believer in unions: "You know, like I say, there's always a few, always a couple that—I believe my husband doesn't believe in unions—but living it and seeing it and knowing what went on—and I saw a lot of the unfairness, I still believe in unions."[3.26]

For its part, the local union celebrated the contributions of its women. Between November 1940 and February 1941, the local's mouthpiece, "Ingraham Union News," regularly reported women in general being "out in front" in "building the union." "At the rate the women are joining the union," the bulletin figured in November, "we will have more women than men in the union."[3.27]

The union in fact needed all the support it could get in the face of a fierce opposition. Many employees rejected the UE, mostly because of continued worker identification with the paternalist Ingrahams. Characteristically, these workers had long ties to the company that transcended gender lines which made them anti-union. Right before the election, some even attempted to promote an independent, employee-run union of "level-headed men or women" and "No Radicals" to prevent a union win.[3.28]

The association between long service records and a disregard for the union assumed particular relevance when applied to skilled men in the movement shop, the most apathetic of all employees to unionism. Watch adjusters, tool and die makers, and machine screw workers collectively looked to their expertise as the basis for achieving breadwinner wages and, in most instances, believed a union could not help them. Moreover, evidence suggests that many of Ingraham's skilled men actually viewed a union as detrimental to their self-worth, believing that a union, in its pursuit of higher wages across the board, obscured the special talents that underlay their wages and their masculinity as well. Watch adjusters and repairers, the highest paid of this group, especially and unanimously rejected the union. John Denehy remembered no union activity among these finishers, who like other non-unionists, often worked at the factory for years. "Maybe [the union got a response] in the other part of the factory, like day workers were low paid . . . and so the hourly rate for a day worker wasn't that great, except in the tool and die room," Denehy recalled. "We had the best tool and die makers you could buy."[3.29]

Despite tool and die makers' high wages, the union did find some support among them as well as from screw machine workers. In both instances, however, support developed independent of the combination of long service and high wages that otherwise impeded unionism. Screw makers rallied behind the UE because of perceived unfair treatment, as Bob Tetro, speaking from management's perspective, related: "The strength of feeling was against a couple of specific supervisors who a significant portion of the rank and file felt were unfair and not fair to them. They listened to and were subject to the blanchments [sic] of union organizers." In the toolmaking department,

Figure 3.3. UE Local 260's letterhead.

the union attracted younger men in the early stages of their apprenticeships. These men saw the UE as a means of more rapid advancement than the regular progression through rigid hierarchies. Against them stood the older men, who, having in many cases achieved breadwinner wages through years of work, opposed the union. In this atmosphere, heated antagonism between the young and old played itself out with unionists reinventing the established male joking practice of nailing work shoes to the floor as malicious behavior against non-unionists. "I took my turn nailing people's [work] shoes to the floor," unionist Carl Kirkby confessed.[3.30]

Still, most men in the toolroom viewed the union as unmanly and an affront to the male-focused paternalism put in place by the owners. In a letter to Edward Ingraham on the day of the election, a mechanic who signed his name as "a satisfied employee," suggested that as "one of your employees of the last twenty or so years" his "thoughts may be of help to you in this matter." Anti-union, the mechanic deplored the union's victory: "I have just learned the results of the election and don't like the results." Assessing its origins, he blamed married women, stating, "It is a well known fact that the married women voted solidly for the CIO." "I don't blame them for that," he followed, "as naturally they will gain by seniority rights."[3.31] Married women, the worker believed, threatened skilled men because they "far outnumber your mechanics and skilled help and by mere force of numbers can dictate to us in any election." "This in my opinion is not a fair deal," he declared, speaking as a breadwinner, "as we believe in the American way of being able to keep our wives home to raise our children." To further underscore the threat posed by married women, the writer suggested that this group lacked the ability to think for itself. In a postscript, he informed Ingraham, "I know that many married women were induced by their husbands (who work in the New Departure) to vote yes." Women not only hurt the rights of skilled men, but they acted as the dupes of their husbands who worked elsewhere, and who, implied the mechanic, should not have a say in Ingraham's affairs.[3.32]

To preserve the integrity of both the firm and its skilled men, the mechanic believed the company had to reduce its employment of married women. Only then could Ingraham defeat the union—with the help of the skilled help: "My idea is for the company to slowly eliminate as far as possible the married women and thereby give us independents a chance to show that we are glad to work for you and believe that you have been trying to do right by us." The mechanic, however, stressed that the firm had to take the initiative, because giving skilled help the upper hand against the union "can be done I am told by the company not hiring any more married women unless they absolutely need the work." "I sincerely believe," he concluded, "I am voicing the thoughts of the large majority of mechanics and skilled workers, and that it will bear fruit." For skilled laborers like this man, the union was feminine by character, had no place in their factory, and married women had to go in order to accomplish this end.[3.33]

Of course, not all married women supported the UE. Some refused to pay the two-dollar monthly membership fee because they did not see themselves as permanent members of the workforce. As a wife, Kay Laviero resisted joining the union, but remembered the campaign from its beginning. "I did not go [to union meetings and rallies]," she said of the late 1930s. "The reason why I did not, I knew that I was not going to work there too much longer and I did not want to join the union. I figured there's no sense in joining the union when I am going to quit working." Even ties to her co-workers, and especially to a chief unionist from her Italian American community, did not overcome her image of herself as a temporary worker. "Everybody would talk and so forth and so on," Kay continued:

because they felt that they wanted me to join, but I did not want to. In fact, the fellow that was the president of the union was a friend of my husband's. . . .

Albert DeCapua, and he was a very good friend of ours. I refused to join and he would say to me, 'But you're not making it easy for us.' I said, 'I'm sorry, but I'm not going to be here much longer. I'm going to retire. No sense in me joining.' You had to pay so much a month and contribute towards the union. I felt, 'I'm going to quit so why should I pay?'

Kay maintained this stand even though she did not leave Ingraham until 1953 to help her husband run his small package store—the reason why she all along saw herself as a temporary worker.[3.34] Despite Kay's case, the reality was that women provided the backbone of support for the UE outside the case shop. The bias of the "satisfied employee" aside, the mechanic was correct to attribute much of the union's support to women.

Figure 3.4. "THE ORGANIZING COMMITTEE of UE Local 260 Bristol, Conn, which is building a powerful local at the plant of the Ingraham Clock Company." *UE News*, December 14, 1940. Case shop worker Anthony DiLorenzo, sitting as president, is third from left in the front row. Anthony DeCapua is beside him, fourth from left.

Women and Status in Locals 260 and 264

The high proportion of women at Ingraham and Sessions guaranteed them visible roles in the UE locals. Women of Ingraham's Local 260 accounted for 11 of the 31 members of the organizing committee in December 1940, and during the campaign leading to the March 1941 election they had a special section titled "Women Today" in the union's bulletin. In the first elections after victory, Mrs. Charles Hartung, representative of the high proportion of married women, won as treasurer, and Jean Porowski, an original negotiating committee member, got elected as assistant to the local's male financial secretary. With the inception of Sessions' UE Local 264 in January 1941, worker Hazel Frateroli sat as both secretary and treasurer and was succeeded in September by Josephine Forcella in the new role of financial secretary. Mrs. Helen Fisher, Forcella's successor as financial secretary, almost ran the local herself while the family owners destroyed it by laying off union members and refusing to rehire them during the process of retooling for war production in June 1942. During the crisis, Fisher took over most aspects of union leadership from a male president, who deferred to her judgment in handling the emergency; she corresponded with UE headquarters as the local spokesperson, often on her own initiative.[3.35]

The leadership roles played by women at Ingraham and the short-lived Local 264 at Sessions, which would not be reunionized until 1956, suggest that women were on equal footing with men. In fact, however, the locals still favored men, in line with both unions nationally and the paternalistic culture of work locally. By 1941 the defense industry had taken off, which provoked a nationwide influx of women into new jobs in industry and made unions like the UE especially responsive to women's needs, compared to the 1930s, when the Depression had focused attention on men because of their roles as breadwinners.[3.36] Bristol's clock and watch industry predated the new trend by giving women a visible position within the union structure, but, as elsewhere, the cultural bias on men allowed them to paint a male face on collective bargaining.

At Ingraham, men dominated the union structure and public displays of solidarity. Bob Tetro did not recall any women unionists at all in the initial period of unionization, although he mentioned a later presence of women. The activists were, he said, "[p]rimarily male" although "I know we had some over the years, some women who were shop stewards and so on who were fairly strong."[3.37] The history of strikes in this early period paralleled Tetro's conclusion. Despite occasional threats to strike, only two actual walkouts happened at Ingraham prior to the 1960s. Both occurred as short sit-down strikes in 1941 following the UE's victory and both established the image of the strike as male. The first in May involved skilled men in the screw machine department while the other, in August, was staged by poorly paid men in the deskilled case shop.[3.38]

In turn, the local union fostered this culture of maleness in failing to educate future unionists about women's critical role for union victory in 1941. George Power, who emerged a leading unionist of IUE Local 260 during the later 1950s, pointed out that by the 1960s women were active in the union, even becoming presidents, as did employee Mary Owens. Well-trained in the lore of his local and the earlier UE, Power presented the later activism as new and representative of a departure from what he implied was earlier apathy. "We had several women presidents," he said of the later period ". . . because you had a lot of women working in the unions." As he explained, "you had the skilled [men on] the screw machine[s] . . . but all the other part of the factory was mostly women, you see, so they [had] become very active in their own rights and in the press room and all over they become strong and got together and they were a strong force because they didn't want the [men to] dominate the whole negotiations. . . . So that's why they become more active in dealing with their problems."[3.39] This image was indeed a departure, contrasting with other female-membership unions like the ILGWU and ACW, where men retained leadership positions. However, it also masks the reality of women's participation in unionism at Ingraham from the very beginning and shows how successfully the UE erased the feminine character of the victory from its history. That the IUE local counted many members from the old UE demonstrates an unbroken continuity in this marginalization from 1941 onward.

Changes at Ingraham and Sessions

For upper management, what mattered was not the marginalization of women, but the permanent place that the union secured in the company. The Ingrahams and their executives pleaded with their workers just prior to the voting to take into account the burdens management shouldered. "We have had difficult problems," the brothers wrote in exacting detail:

> of financing, tariff, planning, production, developing new products, accounting, meeting competitive prices, securing a large sales volume, and so conserving and distributing our income as to pay the highest wages we could to our employees without weakening our position in any of these fields.

The rebuke handed the owner/managers in their efforts "to balance these problems" fostered immediate resentment, which played itself out in the first contract negotiations. "The negotiating committee," declared Bob, "I think, again coming from a rather paternalistic background where we knew what was right for the people to have and to have an Al DeCapua who none of us regarded an individual too highly, [was disgusted]." Bob's own anger increased as he recalled that "the meetings had all the [union] threats, you know the 'Take them out on strike if we don't get what we want,' and so forth and so on." "I'd say management, including myself," he stated defiantly, "resented the fact that that these Goddamned union people came in and were putting their foot in. So it was not something that we welcomed with open arms." Consequently, the contract negotiations, concluded Bob, "were anything but cordial."[3.40]

The owners of both factories let their displeasure be known by restricting worker freedoms. Jerry Nocera remembered at Sessions the hiring of efficiency experts and the resulting caging-in of workers with wire in their departments. Kay Laviero recalled the situation at Ingraham:

> Before the union, it [working at Ingraham] was like a family thing. It was sort of like a big, big family there. If I wanted to get out of where I worked, I would get out of the building and go to the store and buy a sandwich, a cup of coffee or whatever I wanted. I had the privilege of doing that. I don't think everybody could do it, but I was able to do it, but then when the union came in, then the foreman and the superintendents became more strict and they didn't allow you to do a lot of things [like leave work for a break]. We had the union then but before that I had a lot of freedom. Maybe I was an exception, I don't know, but I had a lot of freedom.[3.41]

Although increasingly strict, foremen suffered a loss in status by being forced to balance continued and efficient production with the everyday pressures of union politics. While the union cut into all managerial power at the plant, no one symbolized the loss of that power more tangibly for the workforce than the foremen because they remained the management figures with closest contact to the employees. John Denehy noticed the difference immediately when returning to Ingraham in 1945 after spending the war years at the Waterbury Clock Company. "Well, foremen before were the law," John explained. "[Now] the high chair that they sat on became a little lower [laughs]. Yeah, yeah that big chair wasn't the same as it was before. . ." As a result, union members confronted their foremen more directly by challenging them face to face—affronting the deference demanded by paternalism. Phil Carrier, the longtime assembler on pocket watches, often drew power from union backing to combat perceived injustices. Once, she and the other 14 assemblers on her line refused to do repairs because they were getting defective, crooked screws. The assemblers remained firm on their decision because they had the verbal support of the "head-of-the-union guy." In response, George Tetro, the general foreman of the watch department, fired the entire line. An emboldened Phil went on her own to the superintendent, explained her situation, and got the whole line reinstated. For her, the boss was not so threatening as he may have once been. "We got a union, haven't we?," she reasoned.[3.42]

Faced with diminished power, foremen followed one of two patterns at the plant. Some took a conciliatory stand with their workers, likely because of shared community identities and the need to meet production. Nelson Spring dealt with the UE as a boss on his rise to the position of plant superintendent, coming up from the same backgrounds as many of his workers. He, like other managers, opposed the union, but conceded the loss of power and believed that working with the union proved a better option than fighting it:

> I don't think much about [the union], but I managed to work with it. I mean, it's like everything else, you find a way to work with something or you find a way to fight it and you fight it all the rest of your life, so you might as well find a way to work with it. . . I mean you've got to give somewhere. Somebody's got to give and you can give and make a difference by each giving, sometimes, instead of one giving it all.[3.43]

In contrast, other foremen, namely, in the hotly contested areas of union struggle, took the opposite approach. In the case shop and the automatic screw room, foremen fought aggressively to preserve their positions by taking aim at union members. In the cabinet and sanding department, Oreste DePascale openly cursed activists regularly despite union threats against him until the case shop closed in 1955. He also tried to thwart advancement of union members to higher paying jobs. Joe Grippo, a fellow Italian American and local officer ("a committee man" as Oreste called him) frequently pestered his boss for better treatment, including the chance to make more money. "He wanted this," complained Oreste. "He wanted that." Usually, Grippo cited union support for leverage in making his requests. One day, however, Grippo relied instead on his and Oreste's shared membership in the Italian American community outside the factory. "Are we still friends?" Grippo asked Oreste, who rebuked him outright: "I says, 'In the shop, there's no such things as friends.'"[3.44]

If Oreste believed friendships did not exist in the shop, the family owners at Sessions Clock thought similarly. The Sessionses' feelings manifested themselves fully by June 1942 when they exacted their revenge. Since December 1941 when discussions on a new contract began, the Sessionses showed hostility toward the union's negotiating team, repeatedly stalling the negotiation process. The Sessionses continued to resist signing the new contract into June when the company's retooling for war production gave the family the upper hand against the union. Without a contract, the owner/managers used the temporary factory shutdown and resulting layoffs to eliminate many of the remaining members of Local 264, especially the most active individuals, including Helen Fisher. On June 18 Fisher wrote Julius Emspak, the general secretary-treasurer of the International, that "all of the officers of Local 264 have been laid off with some others and the company through my foreman has said that we won't be called back" but added that "several have been called back but not according to seniority." In a letter in July she closed the book on the local when she told officials that since June "there haven't been anyone there to go on with the Union activities at all."[3.45]

Meanwhile, at Ingraham, the union consolidated its victory during the war by patriotically fulfilling its duty to the United States. Recognition for the contributions of its members occurred during the ceremony for Ingraham's first Army-Navy "E" Award on June 16, 1944. Local 260 played a central role in the event with its president Al DeCapua accepting the award on behalf of all employees. "It is a real token of our Government's esteem of your excellence in the field of production," said DeCapua of his fellow workers. He extended his praise to the larger UE as well. "The members of our local are proud that they too are able to share the enviable record of the United Electrical, Radio and Machine Workers of America with respect to Army and Navy 'E' awards for outstanding achievements in the production of vitally needed implements of war." The UE, he pointed out, had "been honored by these awards more than any other comparable group in the country."[3.46]

UE Local 260, IUE Local 260, and Cold War Ideology at Ingraham, 1945-1951

Troubled times were ahead, however. DeCapua's central role in establishing the union at Ingraham had given him a semimythical status and his actions often-unquestioned legitimacy. This standing helped allow him to weather an early postwar threat from a rival union, the American Watch Workers' Union (AWWU), which in 1946 made an unsuccessful bid to oust Local 260. The AWWU focused exclusively on clock and watches. It aimed, in its words, to unionize "the American Clock Worker," then represented by several different unions, having already successfully organized all of "the employees of the American Jeweled Watch Industry (Hamilton, Elgin, and Waltham)." DeCapua, like the larger UE, had not the knowledge of tariff policy or industry needs that the AWWU championed. Nevertheless, he and three other executive officers dominated an organizational meeting that the union sponsored in March and turned the event into a victory for Local 260, literally laughing AWWU national president Walter W. Cenerazzo and three other union officials off the stage. Bolstering his position within the local, DeCapua immediately studied tariff policy and encouraged International support for higher taxes on imports. He corresponded twice in May with UE spokesman Bruce Waybur on tariffs and delivered Waybur's responses to Dudley Ingraham to prove his, and UE's, commitment to the issue.[3.47]

The president overreached his stature, however, when on January 4, 1947, he filed a portal-to-portal suit against Ingraham on behalf of himself and 18 others from the local, claiming that pay should begin when workers entered the factory gate and not end until they left. In the politically charged climate of the early Cold War that stressed national unity, the action smacked of divisiveness to both the Ingrahams and union members who sided with their employers and fiercely opposed the suit. Self-described "dissenting members" of the local published two separate editorial attacks against the suit in the *Bristol Press* before internal strife within the union caused the withdrawal of the action on January 14. "It's time," admonished worker Grace Lee Kenyon in one attack "we all realized that strikes are just as costly for labor as for capital and that, whether we work or not, living expenses go on and on just the same."[3.48]

Although Kenyon did not refer to the local's president by name, Richard G. Flanagan, a leader of anti-DeCapua forces within the local and author of the second of the two "dissenting" attacks in the *Press* did. Since joining Local 260 in 1946, Flanagan had openly and continuously denigrated DeCapua. A CIO activist in his own right, he had served as president of UAW Local 535 at United Aircraft in nearby Southington from 1943 until that factory closed in 1946. Taking a job at Ingraham as a clock adjuster, Flanagan joined the UE and evidently hoped to ascend quickly to a political position within

Local 260 commensurate with his UAW experience. The entrenchment and popularity of DeCapua and his supporters, however, thwarted the newcomer's goal. An opportunist, Flanagan quickly seized upon DeCapua's controversial move over portal-to-portal pay, unleashing the organizing strategies that he learned at United Aircraft for a dual and self-serving purpose. First, he wanted to oust Local 260's executives from power and, second, altogether replace the controversial UE at Ingraham with a more "American" union that he could direct. A skillful tactician, Flanagan focused on the undemocratic nature of DeCapua's suit and spearheaded the formation of the "Reorganization Committee" within Local 260 to achieve the first of his aims. To announce its first meeting on January 30, Flanagan and his committee distributed a flier aptly headlined "DEMOCRACY 'One for All? Or All for One!!!"—a clear reference to DeCapua.[3.49]

His tactics attracted only a handful of supporters at first, but political developments later gave his group traction. From above, Edward and Dudley Ingraham strengthened their long-standing pro-Americanism after the end of World War II by joining the battle against communism. During his two-year presidency of the Manufacturers' Association of Connecticut (MAC) in 1947-1948, Edward Ingraham led family efforts to stress unity and conformity to American values in the face of the Cold War. In columns for the association's journal *Connecticut Industry*, Ingraham used titles such as the multilayered "Truth Can Keep US Free," and "Danger Signals" to warn of the threat of communism and the need for consensus. Some workers fell into line after Congress passed the Taft-Hartley Act on June 4, 1947. The new law reflected growing Cold War hysteria that regarded the domestic labor movement as a hotbed of radicalism and therefore inherently un-American. The act required unions to file non-communist affidavits but had the broader aim of containing the growth of unionism. In the year to follow, the CIO expelled 11 unions representing more than a million members because of supposed Communist Party domination, with the UE as a prime target. Thereafter, CIO strategy focused almost exclusively on the dismemberment of the UE, whose leaders, DeCapua included, steadfastly resisted taking the oath. The UE was finally expelled on November 2, 1949, and later that month DeCapua took the oath.[3.50]

In the interim Flanagan benefited from the local tide turning against the UE. With fears over un-Americanism increasing during 1947, officials in Hartford-area locals numbered among the first to advocate secession from the UE. By March of 1948 many UE members in Hartford increasingly wanted out of the "red" union and began joining officials in demanding a switch to the more conservative UAW, which had since come out as anticommunist. Auto union leaders quickly won UE Local 265, with over 7,000 members at the Royal Typewriter Company in Hartford, and soon captured almost all of the UE's locals in the armament and typewriter industries in the Connecticut River Valley. By 1949 the Eagle Lock Company's Local 256 in Terryville had defected to the CIO's new IUE. In November the emboldened Flanagan unveiled the "Organizing Committee, IUE-CIO" with about 75 followers and the argument that "there is no place in this Local for the Communist-dominated U. E. and their continued Pro-Soviet policies." On November 21 the self-proclaimed "right wingers" seceded.[3.51]

All along, Flanagan and his group relied on the flimsy charge that DeCapua was a communist. No evidence exists that DeCapua, an avowed Democrat, ever held membership in the Communist Party but during the UE-IUE battle DeCapua's opponents succeeded in creating a lasting impression among some Bristolites that he was a communist stooge. George Power explained, "They thought—everybody thought DeCapua was [a communist], but that was because he was president of the UE and a very strong supporter too." George, who ultimately worked closely with DeCapua in the IUE, highlighted the latter's renowned commitment to unionism:

Of course, I didn't know if he was [communist] or not, but knowing DeCapua as the years went on, he was a real bullhead. I mean, he was involved with the UE. See, . . . he would never give in, even as years went on. When I was involved with him in the union, I had difficulty with him because he was always—'they [managers] were the bad guys and we were the good guys' [is how] he thought of it. No matter what management wanted to do, it was wrong.[3.52]

Amid the paranoia, rumors spread of proof of communist activities within Local 260, causing some steadfast UE supporters to leave the local for the IUE. Carl Kirkby joined IUE because of rumored proof of communism within his former union:

Several of us switched because at the time communism was up front. Everybody was worried about communism and the fact that they sort of proved our union was communist—yeah, they went into the UE office and they got into the vault somehow. I forgot all the details, but there was stuff in the vault involved in taking the communist oath and all that sort of stuff. At least that's what the rumors were. I didn't see this, but that was enough to prove to many of us that we didn't want any part of that.

Carl readily gave into superstition: "So the CIO came along and you're waving the American flag and all. All politics, all politics."[3.53]

Carl's departure, however, also had ethnic overtones. An individual who celebrated his English heritage (and the British nationality of his wife whom he met in Europe during World War II) during his interview, Carl betrayed a bias against the Italian DeCapua. "Well, Al—everybody knew that Al was a communist," he said. "It was one of those things:

Of course, you wouldn't realize it today, but in those

days the Ingrahams were one group and the Italian group were another group. Let's face it. Like in those days, you wouldn't see a pizza, all that stuff advertised on television. Today you'd see it all day long. Back in those days the Italians had their place and the Americans had their place. It was one of those things. Sure has changed since then.

DeCapua, whose parents Louis and Josephine were immigrants, was a foreigner for Carl who sided with the Anglo-Saxon Ingrahams.[3.54]

Other interviewees, both from Ingraham and Sessions Clock, confirmed an increase in ethnic tensions in the late 1940s and early 1950s. Whereas the 1920s and 1930s witnessed interethnic bonds among working families in Bristol's clock and watch industry, accounts of the McCarthy era manifest a tendency toward divisiveness. Agnes Moquin, for example, a Vermont native who worked at Sessions in the early-to-mid-1950s, vividly recalled how one boss, a Southerner named Andy Kahle, "had it in for Polish and French women" because they didn't speak English. She recounted how Kahle would walk through his department, single out these women—many of whom, like Agnes, were recent arrivals to Bristol—and holler at them, "Talk English." Agnes herself betrayed a dislike of Polish women. "The Polish girls," she explained, "worked right through lunch to ruin piecework for the rest [of us]." She and other women resented that this group made 10,000 pieces a day by skipping lunch while the company set quotas at only 1,000 per hour. "They work there and now we can't make ours out," Agnes remembered as the common complaint, adamantly contending that only Poles were to blame. She ultimately argued that the establishment of an IUE local at Sessions in 1956 (after she left) owed its origin to this anger at the Polish women. Clearly, the ethnocentrism exhibited by Carl was part of a larger xenophobic attitude in Bristol during the McCarthy era. For Carl, xenophobic pro-Americanism paid off; his lobbying for the IUE awarded him a position in the fall of 1949 on the Flanagan-backed Organizational Committee.[3.55]

In light of the rampant xenophobia, even officials in the local Catholic churches leapt into the dispute to prove the Americanism of their multiethnic parishes. In the days leading to the secession at Ingraham, four churches—three in Bristol and one in Forestville—weighed in, urging the local's largely Polish and Italian members to support the new IUE. At two of the Bristol churches, St. Ann's and St. Anthony's, clerics told union members directly to support IUE. Father Vincent T. Iannetta, assistant pastor of St. Anthony's—where DeCapua himself worshipped—led the pro-IUE stand, arguing that Catholicism and Communism were incompatible and pointing out that the "church would never have interfered if the question was concerned merely with the formation of a labor union."[3.56]

As the crisis entered the late winter of 1949-1950, the IUE group could not count on the Yankee owner/mangers for support. The Ingrahams likely preferred the right-wingers to the UE, whose presence had long vexed them, and probably thrilled privately at the thought of a UE failure. Openly, however, the owners refused to get involved owing to their punctilious legalism. Without managerial support and denied the ability to label DeCapua a communist, the secessionists again rethought their strategy as January waned. They settled on raising the old torch of communism but this time directing it at the international UE. Their persistence secured an election from the NLRB on May 11, 1950.[3.57]

Up until the election, the battle underscored the gendering of public shows of solidarity as male by assuming

Figure 3.5. *Clockwise* announcing NLRB election against the "IMITATION Union (IUE-CIO)," May 10, 1950. E. INGRAHAM COMPANY PAPERS.

a decidedly masculine character. Given the Cold War emphasis on traditional family values of the man as the public figure and women as creatures of the private sphere, the struggle had shifted focus away from the many women members. The rules of battle fixed attention instead on UE president Albert DeCapua and the IUE's leader Richard Flanagan. Only three days before the May 1950 election between the UE and the IUE did the UE, in a fight for its life, actively rally women against the "imitation union" in a special meeting for all women workers. Women's support of the old UE helped to keep the local afloat on election day: the union won 797 votes, which the IUE outpolled with 888 but which also prevented the usurper union from getting the 924 votes required to win the election. The NLRB scheduled a runoff for June 8, in anticipation of which the UE local took to the airwaves at the end of May with a roundtable discussion with two women about working women's seniority and wages.[3.58]

Figure 3.6. 1950 contract between Ingraham and IUE Local 260.

Feeling that it won the election fairly, the IUE chafed at the postponement of the runoff. Its members' seething anger boiled over into open hostility on June 7 when they interrupted a planned noonday meeting at the shop gate by Local 260 and tied up all the parking space across the street from the plant to prevent easy access for the UE members. Tensions on both sides mounted; the UE was furious over the interruption, especially since the IUE had recruited outside members to help to ensure confusion. Carl Kirkby, who participated for the IUE, recalled the moment vividly. "Well, one day we were going to have the election, trying to get the new union in and the old union out." He described the scene:

> All along this side of the sidewalk on the factory side, stood all the good old UE employees or whatever and on the other side were the CIO people with their goon squads, and I mean there was actually a goon squad there. . . . The people were imported from—well, they were saying from New York and Hartford and all the big cities, you know, looking for a confrontation and the police were in between us to keep us apart. I often wonder what would happen if the police hadn't been there. It would have been a slaughter[.]

Tempers were indeed heated: "You'd seen this in the past but you'd think it could never happen here it was right in front of you. So, you were living it [and] you expected something to happen any minute because everybody was hot. It was a very hot issue kicking one union out and

Figure 3.7, right. October 1952 issue of IUE Local 260's *Tick Tock*, which replaced *Clockwise*. E. INGRAHAM COMPANY PAPERS.

bringing another one in." Still, officials reported the next day's voting as "orderly." The final tally was IUE 1059 and UE 779, with the NLRB declaring 35 ballots void. Since the earlier vote, the UE had lost 18 supporters while the IUE-CIO gained 161. The defeat devastated UE Local 260 so gravely that not even a statement of concession survives.[3.59]

Elated, IUE Local 260 members rallied around Richard Flanagan, who now held his long-desired post of union president. Knowing that so many Ingraham workers opposed them, however, the victors worked quickly to stave off rebellion and legitimate their hold on power. Taking the IUE–CIO line of working amicably with management, Flanagan and his executive board immediately entered into friendly negotiations to replace the UE contract that had expired on June 1. Their willingness to accommodate management won rapid approval of vacation pay for 1950, a moot point since it had not been in jeopardy, but which the Ingrahams themselves, likely pleased at the UE defeat, publicized in the factory. That same month the IUE introduced its monthly publication entitled *Tick Tock* to replace the UE's *Clockwise* and win new recruits. In September the IUE won another victo-

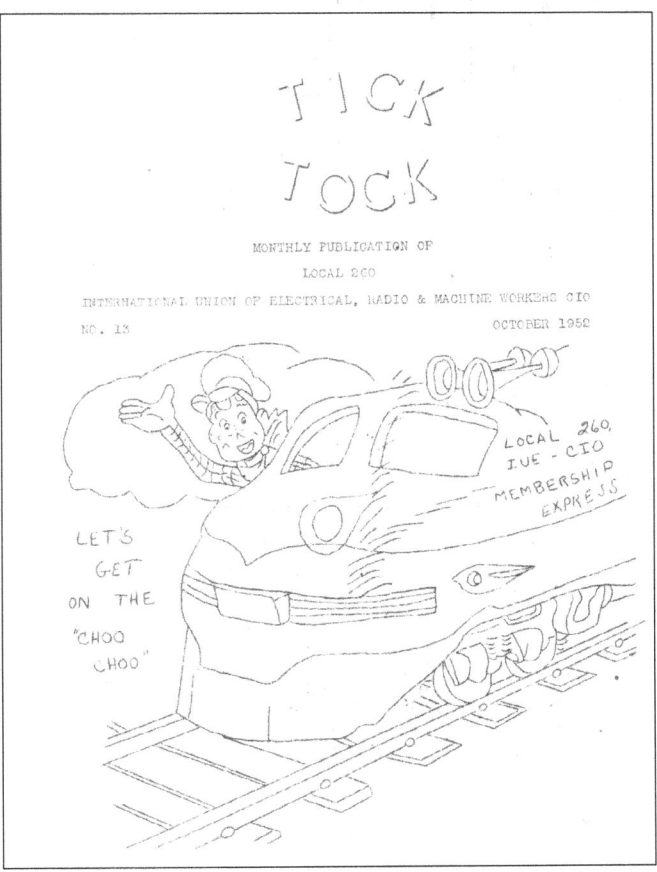

ry in its efforts at consolidation by signing with Ingraham a five-year contract that, among other stipulations, gave all employees an across-the-board increase of seven cents per hour. Edward Ingraham seemed content at the lengthy duration of the contract, believing that management could take advantage of the IUE paranoia against remaining UE elements and keep the latter in check. The IUE had reason to be "scared to death of the UE group," as Ingraham put it in his memoirs. On the surface, workers had accommodated themselves to national developments to avoid persecution. Moreover, the continued rise of Senator McCarthy enhanced the ideological reign of terror, thwarting open UE attacks. However, the IUE suffered multiple defections that diminished the ranks of its membership. The outflow continued at least through June 1951 when the local lost another four more employees.[3.60]

By mid-1951, however, the IUE had rooted itself firmly at Ingraham and helped to ensure the nationwide union schism in the electrical industry. The battle at Ingraham left division and distrust among unionists in its wake. George Power remembered intense factionalism, especially regarding Albert DeCapua, the vanquished symbol of the UE. George recollected that "as years went by he [Al] changed some, but he went down very hard on that [defeat]. He fought for the UE and then when he got beat, he wouldn't have nothing to do with anything for a long time." DeCapua eventually recovered from the defeat and joined the IUE, working closely with the young George and showing him the ropes of unionism. From 1958 on, he served multiple terms as Local 260's president, succeeding his nemesis Flanagan. Even after rehabilitating himself in the IUE, however, he refused to forsake the principles he learned and practiced in the UE, and in effect used the IUE structure against itself to preserve those values. Pat Letizia, a worker at Ingraham first during DeCapua's UE days and later again during his IUE presidency, confirmed this reality. "He was a really good president," she declared unequivocally, "because he was for the people." Refusing to adopt the IUE's collaborationist unionism, DeCapua embodied the principles of the former "rank and file union" as the UE called itself. That Ingraham's workers ultimately championed him speaks to their own preservation of this unionism, even if bending to national anti-communism and forsaking some of its fire.[3.61]

Chapter 4

Tempus Fugit: Livelihoods in Trouble

On July 1, 1960, 200 skilled men staged a wildcat strike at the "Ingraham Company" as it was now called. They protested the decision by the firm's non-family management, then four years old, to relocate the factory outside of Bristol and build a smaller plant nearby. A year before, Ingraham opened a new assembly plant in Laurinburg, NC, which took jobs away from Bristol and suggested that local workers' livelihoods were unsafe. Since then, the company's president, Robert Cooper, made plans to move ahead with Ingraham's relocation outside of the city. Skilled men fumed and unleashed their anger while the company held its annual two-week summer vacation after a rash of layoffs. Peaceful pickets, now with the IUE Local 260's support, continued until June 19 when a skirmish occurred between unidentified strikers and police. The incident resulted in slight injuries to Mrs. Eleanor Casey, the local's recording secretary, when pushed against a fence, although the *Bristol Press* reported no arrests. The strike ended on the 25th with the tentative approval of a new two-year contract, the factory secure in Bristol, and the promise that workers were free to resume their former jobs without prejudice and with all of the rights held prior to the strike—plus an across-the-board raise of five cents per hour as an added bonus.[4.1]

The strike reflected the limits of paternalism amid industrial decline. Only a few years earlier, the Ingrahams and Sessionses let control of their factories slip into outside hands, with the latter selling their business altogether. Unhappy with the decline of their industry amid unfavorable government policies and business management decisions, the families stepped back as new managers and owners outsourced jobs in the search for a cheaper workforce and greater profits. Outwardly, Bristol looked stable, having grown steadily since the end of World War II: suburbanization from Hartford had increased the population of the city from 35,961 inhabitants in 1950 to 45,499 in 1960 and 55,487 in 1970.[4.2] Economic stagnation, however, replaced the city's traditional industrialized culture, especially in the clockmaking sector. Managers introduced new timepiece items in the effort to remain competitive. Ingraham even returned to making fuses, a key product that sustained the company through the early 1970s. At the same time, however, dwindling profitability caused the firms to explore new sources of labor, as symbolized in the opening of the Laurinburg plant. In opting to move production outside of Bristol, businessmen left locals with an ever-decreasing job base. Ultimately, even Sessions showed the tensions of job insecurity in a costly nine-week strike in 1967. Sessions closed in 1970, and Ingraham remained opened under its name until 1974. All the while, workers made choices and adapted new identities in defense of their selves, families, and communities in the transition to a postindustrial society in which Ingraham and Sessions no longer played a part.

Sessions, Ingraham, and Changes in Management

By the mid-1950s declining sales, coupled with wages rising faster than profit, crippled Sessions Clock financially and sent the family owners scrambling to keep the company afloat. Sessions required a mortgage loan of over $1.2 million to continue operation by 1956 and saw its employment shrink to 100 by early 1957. One creditor, Consolidated Electronics Industries Corporation of New York, saved the company with another loan for $400,000. The Sessionses responded with a deal allowing Consolidated to purchase controlling interest in the company on April 1, 1958. By terms of this agreement, Consolidated Vice-President Arthur W. Haydon, who started his own business in leased space of the factory during the Depression, became president while W. K. Sessions Sr. remained chairman of the board. Haydon immediately brought in additional lines, namely, special timers and industrial motors, which built upon an extensive advertising campaign he instituted in September 1957. In 1958 he added the profitable MEPCO Division for the production of precision wire-bound and carbon resistors for the larger company of the same name. Sessions initially stabilized, with sales increasing regularly after 1957 up through 1966. Nonetheless, strides in sales failed to generate sustained corresponding increases in profit for the fundamentally unstable company. Symbolic of the situation, relatively small changes in sales produced dramatic fluctuations in profitability. Sessions also suffered from tax problems, as had happened during the 1920s. Bristol regularly attached city tax liens to the factory from the mid-1950s onward, with the firm owing $195,000 by 1960 alone. All the while, the workforce steadily dropped, falling from 809 in 1955 to 110 by 1967.[4.3]

Sessions owed primary blame for renewed instability on its failure to produce popular clocks. Horologist Chris Bailey has correctly dismissed the Sessions line in latter years as "at best cheaply made and unimaginative." The clocks had only moderate appeal, a problem compounded by poor decision making by management. The company offered a line of decorator clocks that used spring-driven lever movements, which negated the need for an unsightly service cord and proved popular with consumers. The firm discontinued the models after introducing a new battery-driven unit, meant to surpass the key-wound movements in desirability by not requiring winding. Antiquated engineering at the firm, an impediment that the new owners did not overcome, made the new movements easy prey for superior foreign models that relied on evolving quartz technology. The revamped line ended dismally as a sales disaster. Meanwhile, hoping to target sports-oriented consumers, Sessions introduced football clocks, sanctioned by the NFL in 1965. The buyer could personalize these models to his/her favorite team by choosing as the face one of the included self-adhesive stickers from all 14 clubs. Only modest sales resulted.[4.4]

One of the most unsuccessful of the later clocks proved arguably to be among Sessions' most innovatively designed clocks.

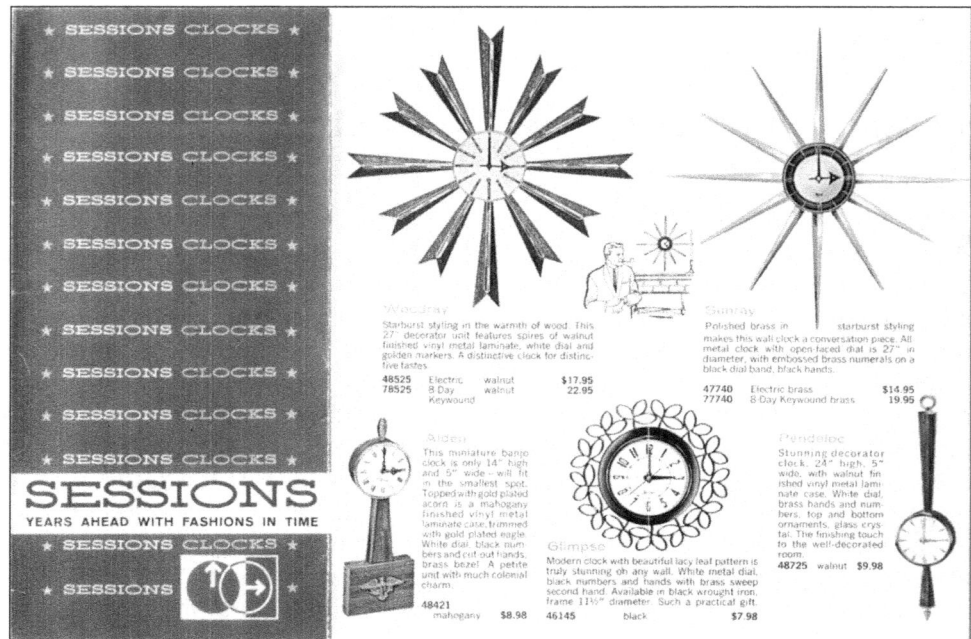

Figure 4.1. Sessions clocks of the 1960s.

In July 1965 the firm introduced a family-planning clock built into an existing female-oriented model called the "Lady Sessions." Invented by Los Angeles lawyer Maurice Gordon, the clock incorporated a mechanism that when set properly with a woman's menstrual cycle indicated days of greatest and least fertility according to the rhythm method of birth control. Office employee Domenick Dellario recalled that the clock drew immediate attention, in particular from South American governments that hoped to use the device as birth (and population) control for largely Catholic populaces that in concert with Rome rejected other contraceptives. The plan backfired. To function, the clock required constant and steady power; any disruptions in its electrical supply necessitated a complete resetting of the planning mechanism, a problem because the clock found its primary uses in usually underdeveloped areas that lacked steady supplies of electrical current. Targeted populations often did not bother or know to reset the clock whenever its power source failed, rendering the planning mechanism invalid. To further undermine its legitimacy, the clock, as a contraceptive device, drew the ire of the Catholic Church. Domenick said the Vatican even issued a papal condemnation against the invention.[4.5]

In contrast to Sessions, Ingraham in the 1950s thought that continued viability meant turning toward outside management while retaining ownership. The company, former Executive Vice-President Bob Tetro explained, searched for an executive with a background in marketing to confront the problems of a declining domestic market and settled on Chicago native Robert Cooper. Under the arrangement, Dudley and Edward Ingraham retired to the board of directors, with the latter retaining his position of chairman of the board. To signify the reorganization of management, the firm changed its name to "Ingraham Company," dropping the initial "E" from its letterhead. The switch to an outside president, however, did not go as smoothly as the change in name. Cooper, who lacked the connection to Ingraham that Haydon had to Sessions, immediately upset the longstanding distribution of power that the Ingrahams achieved through generations of management. He stripped his managers of the right to chart Ingraham's future, choosing instead an outside source. Cooper affronted his subordinates in their day-to-day decisions about running the factory, causing almost instantaneous friction for managers accustomed to the Ingrahams' "laissez faire" style. Bob Tetro, the second generation of his family at Ingraham, tried to lobby on behalf of middle management, intervention that cost him his job in 1957. The hatchet fell on other managerial figures as well. Labeling them poor managers, Cooper in 1959 started removing old stalwarts whom he believed stood in the way of progress. He banished Nelson Spring, the former production superintendent, to the new plant in Laurinburg, telling the longtime employee, then in his middle fifties, to comply or be fired. Lower level figures were less fortunate; longtime bosses, such as Al Calderoni, a low-paying watch worker who rose to foreman of the electric clock department, lost their jobs.[4.6]

Cooper also tried to update Ingraham in other ways with limited success. He introduced a new line of electric clocks, including 31 kitchen, wall, and desk models, and then unveiled a new compact 8-day clock movement that proved widely popular in a variety of novelty timepieces and decorator clocks.[4.7] This spring movement competed well with other makes, including the Sessions version, and after the opening of the Laurinburg factory, constituted the sole product for which the Bristol plant

Figure 4.2. Ingraham President Robert Cooper, second from right, front row, 1957. In first row at left is Bob Tetro. Dudley Ingraham is second from right in top row, his son Seymour is third from right.

both made parts and assembled completely. Cooper also discontinued the company's increasingly unprofitable wristwatches in 1959. Moreover, he advanced Ingraham's growing reliance on the production of military fuses by regularly taking on government contracts that kept the company afloat.

Despite these measures, Cooper did not increase profit at the ailing company but instead lost money. "Every year under the management (or better put "mismanagement") of Robert E. Cooper," foamed chairman of the board Edward Ingraham in his personal memoirs, "the company lost money and lost ground until Cooper was fired on July 3, 1961[.]" Two noncontroversial presidents followed, Bret Neece (1961-1963), former chairman of the board at New Britain's Landers, Frary, and Clark, and then businessman Wesley S. Songer, whose administration is most

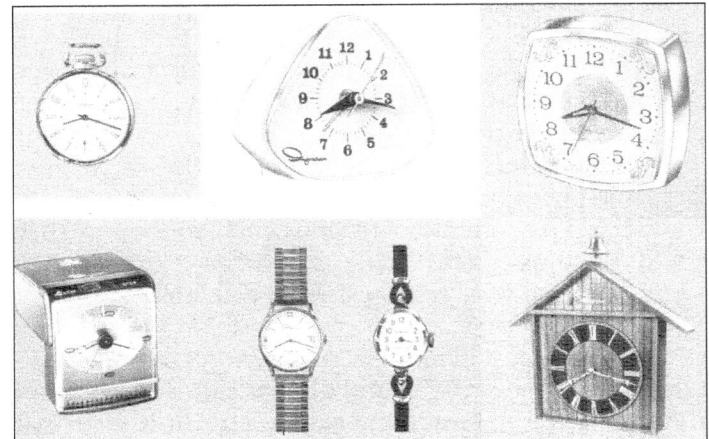

Figure 4.3. A sample of Ingraham's catalog from the mid-1960s.

Figure 4.4. Ingraham's new home on Redstone Hill Rd., Bristol, mid-1960s.

notable for the 1964 move of Ingraham into a smaller one-floor factory on Redstone Hill Road in East Bristol near Forestville. Songer lasted until March 1974 with the reorganization of Ingraham Industries by McGraw-Edison, which purchased the company from the family owners in 1967. Thereafter, Joseph A. Bussmann, of the corporation's Controls Group, headed the company, now renamed as Bussmann Division of McGraw-Edison and thus ending the history of the firm under "Ingraham."[4.8]

During his troubled tenure, Robert Cooper made one additional change at the Bristol factory to boost profit, a maneuver that prefigured the move to Redstone Hill Road and symbolized the president's legacy for his successors: a reduction of the local workforce. As Cooper downsized the workforce, he looked southward for a cheaper source of labor to assemble Bristol-made parts. With the board of directors—the family owners included—behind him, Cooper opened the Laurinburg factory in 1959 in what the firm saw as a highly competitive move. On many levels, Laurinburg, a non-union city of 12,000 in rural South Central North Carolina, was a logical choice to exploit for the proposed factory, one-sixth the size of the Bristol equivalent. As Edward Ingraham wrote in his memoirs, "local citizens agreed to put up over $600,000 to build a new 100,000 square foot plant to lease to us on a basis which provided for our ownership of the plant after 20 years." The new plant, however, failed to turn a profit during its early existence, Ingraham continued, " . . . as there was no skilled help in the area capable of making and maintaining dies, set up jobs, maintenance

Figure 4.5. Ingraham's controversial assembly plant in Laurinburg, NC, mid-1960s.

thereof, dies and set-ups of fine subassemblies, screw machine operators, etc." To counter this hindrance and get the plant going, "Many supervisors were sent down from Bristol as well as the Manager."[4.9]

What this group found, however, proved equally disheartening. "I thought we were going to have a place, but there was no damn factory there," recalled Nelson Spring. "It was just a piece of land [laughs]." As manager, Nelson quickly solved this issue. "And to start off with," he explained, "I had to get something going so I found an old furniture store that had an attic that was partly filled with furniture and junk and everything else." After cleaning it, he set up his operation:

> I started one girl in an attic of a furniture store, . . . and I had to show her . . . how to assemble a watch. . . Then after I got her so she could do it quite regularly, so then I took another girl and brought her up and let her start teaching the other girl, and by that time she kept improving, which made it better for me all the time because I kept using her as a teacher.

To ensure that this system took hold, Nelson recruited seasoned watch workers from Bristol to relocate permanently to Laurinburg; as of 2002, Bristol native Sally Reimer still lived there, having been joined by her sister Ida Fisher, who arrived after retiring from McGraw-Edison and died there at age 91 in 2001. Even with the Bristol connection, however, the Laurinburg plant suffered from yet another setback, one that related back to Bristol. The home factory failed to supply an abundance of parts for assembly, requiring an even greater emphasis on the experienced Bristol workers to keep the 500-employee factory afloat until the fall of 1961 when Cooper's successor Bret Neece achieved balance.[4.10]

Worker Responses to Laurinburg at Home

"With Cooper in the saddle in Bristol and morale at low ebb," Edward Ingraham wrote, "little could be expected of Laurinburg." A rumor quickly circulated that the North Carolina operation would rob Ingraham-Bristol of supposedly two-thirds of its more than 1,000 jobs. This unsubstantiated gossip, which even the *Press* published as truth, heightened the sense of alarm. IUE Local 260, which had spent much of the decade benignly supporting management, openly opposed Cooper and won the support of Mayor James Casey. Casey lobbied for federal dollars to appease upper management in defense of local jobs, turning to state senators Prescott Bush and Thomas Dodd. An onslaught of defense contracts in 1959 and 1960 followed and outwardly quelled passions. Still, Cooper looked for ways to downsize Bristol, leading to the July 1960 strike. In addition to other outcomes, new Mayor Walter Murphy used the incident to secure the company a defense contract worth 3.05 million dollars from the government, which helped to guard job security.[4.11]

Job security also meant abandoning the old factory on North Main Street, too large and unprofitable to sustain the reduced workforce of 1960. In 1964 employees moved to the factory on Redstone Hill Road. The new workplace, popularly known among workers as "Redstone Hill," left various impressions upon workers accustomed to the old, six-floor factory centrally located in Bristol. Walantyna Sakowski had mixed feelings, not approving of the austere, modern factory and in particular the lack of ventilation, which caused temperature extremes for the workforce. "We were cold in the winter and hot in the summer," Walantyna deplored. "So there was no happy in between." The wintertime coldness caused her to start grudgingly wearing slacks, attire that she considered unwomanly but nonetheless necessary. Walantyna, however, didn't mind the new location, in large part because she lived on Redstone Hill Road. Whatever their opinion, workers quickly adjusted to the new environment, which housed the operations of Ingraham and its successors in Bristol until closing as Bussmann's on April 1, 1991. At that point, many workers from the original factory still remained. As late as March 1986, Bussmann employed 34 men and women who started at Ingraham in 1960 or before, with 12 of this number beginning prior to 1948.[4.12]

Fuses As Ingraham's Salvation

The profitability that underlay the success of Redstone Hill under the Ingraham name, and even the ability of the company to continue through the 1960s, emanated from company participation and ultimate reliance on what Dwight Eisenhower at his retirement as U.S. president termed the all-powerful "military-industrial complex." As imports and low tariffs destroyed the domestic clock and watch industry, Ingraham turned more and more to fuses, which breathed renewed life into the dwindling company even when outward signs suggested inevitable demise. Both Ingraham and Sessions made antiaircraft fuses for the government during the Korean War until the curtailment of defense contracts in June 1954. While Sessions periodically got contracts through the 1950s, Ingraham became dependent upon them from 1957 onward. From then forward, the company prioritized the fulfillment of government contracts for fuses and so-called "boosters," or supplemental fuses, above all else. By 1974, however, the clock had run down for Ingraham; a declining need for ordnance as the Vietnam War brought an end to the company under the Ingraham name.[4.13]

Ingraham employees made a smooth transition from clockmaking to fuse manufacture, and many floor workers did not pay attention to the switch so long as they could make as much money as possible. Women performed the bulk of the presswork, assembly, and floor testing of fuses and boosters and made up the majority of the workforce. Lillian Rock countersunk fuse plates after going to work in 1953 and did not think much about the switch. "[We] were countersinking the little plate that I think went into some kind of fuse," explained Lillian, "but at the time I didn't know what I was doing and I didn't care." Lillian could recall only two other details, unrelated to recognizing the defense character of the job but offering telling insight into work cultures of the early 1950s: "I remember that it was in the front of the building so that I could see out to the street. It was a long room, quite long, benches going up the aisle on either side, and everyone could have their own radio." For Lillian, the work she did mattered more than the ultimate purpose of the product she made.[4.14]

At least one floor worker, however, appeared more conscious of the military nature of her jobs. Georgia Parlyak, a resident of closeby Unionville, CT, went to Ingraham through her friend, longtime employee Marge Dziedzic, in 1959 or 1960. The 28-year-old married mother worked only about a year before being laid off and then returned on a call back before leaving permanently in May 1962. In both instances, Georgia labored on fuses. Starting on the day shift, Georgia trued fuse hairsprings, which she found boring and sleep inducing because of her household responsibilities. "I did what they called a truing of a spring that went into I think a bomb [laughs]," Georgia said:

> I don't remember now. But, of course, at that time [I] was not a morning person. And it really was hysterical because, you know, you had this little disk and you would have to put this spring in. And these tweezers, long tweezers, that you would have to turn these and get these to aim, you know, in a certain direction. And you had to look like through a little spyglass, you know. And so, you know, here I had a few children that you're up half the night with, so that you know you don't get a lot of sleep. You know how it is. And so I'd be doing this and [nodding off] because it's, you know, very tedious work[.]

Georgia did find stimulation through her inclusion in the female networks at the factory. She enjoyed solidarity with other women, many of whom were mothers like herself, and, like workers in earlier periods, shared with them moments of laughter that broke the monotony.[4.15]

However boring she found truing hairsprings, Georgia felt fortunate because of her initial experience which involved handling live ammunition in the loading room. Without a counterpart in traditional clockmaking, the loading room represented a distinctly military sphere within the company. As such, it alerted Georgia to the warlike character of fuse products, disturbing her in the process. For Georgia, who stressed in her interview that she based her womanly worth on her role as a wife and mother, having to load fuse charges proved incompatible with her implicit belief that women did not handle ammunition. "I had to put in these bullets," she remembered, "in a certain capsule or whatever." Georgia could not stomach the work and asked for a transfer, which her bosses evidently granted:

> I really lasted only a couple of days because I was so nervous. Every time I put it in it would seem like it didn't [fit], whatever I was doing wasn't right because the next thing I knew all the bullets were on the floor and I'm going, 'I'm having a nervous breakdown,' because, you know, I really had never held a gun or, you know, did anything of that kind, you know, that you would work with ammunition or so forth. . . . Live ammunition. So, I said, 'I think you better take me out of here.' And so, I guess they agreed because that's when they moved me. And I think that's when I went to that other little job. . . . I don't remember what they called it. But you were to adjust the . . . mechanisms for, you know, grenades [sic].

Georgia's relief at her removal from the loading room was itself only relative; she jumped at the opportunity to leave Ingraham in May 1962 and waitress at a Howard Johnson restaurant, a job that she held until 1985. In contrast with her career at Ingraham, Georgia expressed real satisfaction with her new job, which she proudly said mirrored her first employment at running a Wool-

Figure 4.6. Assembling fuses at Ingraham, mid-1960s.

Figure 4.7. Fuse-making machinery at Ingraham, 1970.

worth's counter in 1948. She cherished these service jobs because, for her, they paralleled household responsibilities from which she derived her identity as a woman, something she could not say about fusemaking.[4.16]

While happy to leave Ingraham, Georgia acknowledged that her various jobs on fuses imparted her new skills, from which she could have benefited had she remained in the industrial sector. She especially enjoyed working with new electrical devices used to test the variance of fuse mechanisms. Georgia performed this high-tech work during a stint on kick presses:

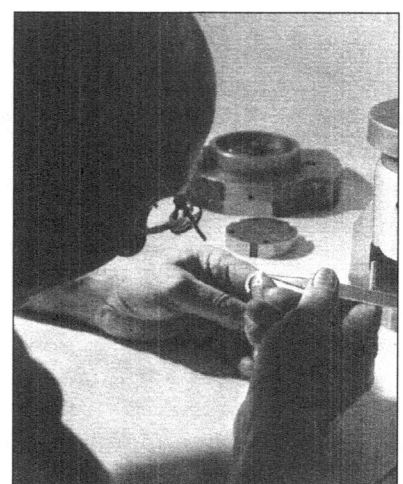

Figure 4.8. Toolmaking at Ingraham, mid-1960s.

> The only thing that I remember is I decided I wanted to try this new area that [where] they . . . [inspected] the little grenades when they were all set, I believe, but I believe it's before they have all the powder and bullets or whatever they put in there. But I worked on the mechanism part and what, they had a little machine and you would have to [adjust to] what the machine registered. It was like a computer in a way because it would show you the screen. And then it had to be a certain variance so you would have to do some adjustments on it. All I remember is working with a screwdriver on that.

Georgia's words paint the application of technology to work as gendered. Indeed, in this instance, management assigned electronic variance testing to women. Untainted by what Georgia saw as the unwomanly aspect of live ammunition, this job appealed to her with its sex-typed distinction. Beyond discussing her exposure to new technology, Georgia expressed affinity for the work. "But it was a nice job," she gushed. "It was a clean job. It was a fun job really. I really liked the job a lot."[4.17]

Still, the possibilities for advancement were limited. Women like Georgia learned to read monitors that suited their employment in the evolving economy but, as floor workers, they did the same deskilled jobs as in clock and watch manufacturing with little chance for upward mobility. In contrast, white-collar engineers, whose designs lay the foundations for the success of the fuse program, enjoyed often rapid advancement. At the forefront of fuse production, engineers spent increasing amounts of time traveling across the United States to coordinate with other manufacturers of ordnance. As such, the definition of their paid labor expanded the working identities of engineers beyond the traditional confines of Bristol. In so doing, it distanced them from the working class of which they remained a part.

Engineers and the Fuse Program

The history of engineers in industry at Ingraham and the larger United States demonstrated pronounced change with the advent of the military-industrial complex. Between 1900 and 1940, Ingraham employed only a handful of engineers and did not even have a formal department, corresponding to the inappreciable size of the national profession during this period.

The explosive growth of the defense industry during World War II and the Cold War, however, changed the size of the profession both nationally and locally. Between 1941 and 1964, employment in aircraft, missile, and rocket manufacture—the backbones of the munitions industry—grew to exceed that of automakers, the nation's traditionally largest employers. As in the case of Ingraham, the air defense industry in turn relied on sections of the electric industry, including the clock and watch sector. In both instances, employers staked continued growth on engineers, who, numbering 872,000 nationwide in 1960, served as the professionals of the technical-scientific revolution that underlay the national defense buildup. In the electronics industry alone, engineers spurred employment to 700,000 in 1964 by providing the designs for government ordnance which helped to drive the sector's postwar expansion.[4.18]

A disproportionate emphasis on engineers resulted in both the air defense and electronics industry that created a group imbalance against other professions, discernible even at the time. At Ingraham, this imbalance manifested itself in both the creation of a specific department and employment of some 20 engineers by the 1960s, most of whom had degrees from not only high school but local technical schools as well (a college degree was not yet a professional necessity). These developments represented a departure in employment patterns at the factory; whereas other departments shrank or closed altogether in downsizing that occurred in the latter 1950s, engineering took hold and expanded, at least in relative terms, and did so with educational requirements not necessary elsewhere in the plant. After 1960 the work environment of their own department reflected the nonmanual character of engineering, distinct from other areas both in terms of clinical cleanliness and employee attire that called for shirt and tie. Likewise, engineers had among the highest pay scales in the factory, replacing now-defunct timepiece adjusters who formerly ranked as the best paid (although toolmakers and screw machine operators continued to command comparatively strong wages).[4.19]

As well, engineers realized that their job classification afforded them real bargaining power in making career moves. In 1958 "Bill V.," an anonymous engineer with 11 years of service, suffered a layoff from research and development in the tool room following the termination of a fuse contract. Through connections, he got a job for

Figure 4.9. Ingraham's engineers at work, mid-1960s.

nine months in Winsted, CT, at the Gilbert Clock Company at a 50 cent an hour pay cut from what he had at Ingraham, thankful all the same because, having already begun building their first house, he and his wife had acquired heavy debt. In the interim, a renewal of fusework at Ingraham made Bill's old foreman eager to rehire him. Ingraham lured him back with a 20 cents an hour raise from when he left, a circumstance that greatly relieved Bill's financial stress.[4.20]

Collectively, these changes in lifestyle fostered a consciousness that increasingly approached middle-class status. Nevertheless, engineers remained very much part of the working class. Their continued remuneration through an hourly wage and not the salaries that accompanied middle-class occupations confirms this reality and reminded Bill of the limits of his job as well. The physical boundaries of that work also masked the effect of a larger trend toward slowing advances in payscales and benefits plus decreases in overall employment numbers across the board. Concurrent with engineers, the nationwide growth of white-collar jobs rapidly accelerated in the 20-year period after World War II. This circumstance may have seemed economically and politically desirable in the Cold War climate of the 1950s when dominant ideology demanded conformity to middle-class norms. However, Stanley Ruttenberg, then research director for the AFL-CIO, argued otherwise. "This shift to white-collar occupations is a move toward lower-paying jobs," Ruttenberg warned in 1959. "It has produced and will continue to produce a period in which employment will grow at a decreasing rate and wages and compensation will increase at a much lower rate."[4.21]

The example of Ingraham followed this pattern precisely. Not only did the workforce decrease from the 1950s onward, pay and benefits did not advance as proportionate to earlier periods. Instead, this compensation more or less stabilized, increasing only incrementally at various periods and making Ingraham an acceptable choice for work in the minds of Bristol's working class but no longer "the" place to earn wages as it had been when a high-paying producer of clocks and watches. Even engineers themselves did not benefit in the long run. Whereas they held admirable pay, the earlier job of adjusting offered better economic remuneration in its day. Under such circumstances, the notion of engineers progressing materially in the shift from blue-collar to white-collar labor stood on shaky ground.

Despite this fact, engineers had one aspect of their job description that clearly separated them from other workers: they had to travel either to coordinate their efforts with other munitions makers or test lots on government proving grounds. Traveling represented a change in working conditions that created a distinct identity for engineers, since they alone among workers left the factory as part of their job. As a consequence, descriptions of travel figure heavily into former engineers' interviews.

August "Augie" Erling spoke with particular pride of his job in engineering and how he progressed through the ranks. Augie began in drafting on clocks and watches in 1948. He climbed rapidly with the demand for fuses created during the Korean War because, as he explained, "then they had an opportunity to get a prime contract for manufacturing mechanical time fuses for the government, and so at that time we had the government drawings on hand and we had to convert to our own Ingraham drawings." Ingraham needed Augie's specialized services for those blueprints. "So that if we ever had to make changes or something of that sort," Augie ex-

Figure 4.10. Conducting tests in the model shop of Ingraham's Engineering Department, mid-1960s.

plained, "we'd have our own drawings to work with, and I spent a lot of time on those Ingraham drawings for the government fuses, and I learned to know the fuse very well just by doing that." This reliance on Augie led to a promotion to what he termed "a junior engineer" on fuse work, which dominated his career until he was let go in the mid-1970s.[4.22]

Augie without solicitation turned to the subject of traveling, a circumstance that dictated his career by the early 1950s "as a contact man with a lot of government arsenals throughout the country." The job required Augie to visit testing grounds regularly:

> I'd do a lot of traveling for the company and when we did test firing on the fuses, I spent a lot of time going to the proving grounds [at Aberdeen, Maryland] and observing and making notes for the company, bringing back information so that we could improve on what we were doing and if we had troubles, we had to work on it to try to straighten it out, see where the problems were and have them corrected.

By the late 1950s Augie did less traveling, the result of a declining need for further modifications in the Ingraham designs then being manufactured. The outbreak of the Vietnam War, which corresponded with renewed government demands for improving the current fuse design, put Augie "back on the road once again." He remained on the road so long as military orders for fuses continued.[4.23]

Like Augie, Bill V. chose to illustrate his job by stressing the travel aspect of his work. After his return to Ingraham in 1959, Bill took the position of lab technician, which he held until 1974 when he, like Augie, lost his job due to declining demands for munitions. In this latter period, Bill served as the company's witness at ballistic tests. "We used to have proving grounds," Bill explained:

> When we made a lot, which at that time was 5,000 pieces, we took a sample for ballistic testing at Jefferson Proving Grounds in Madison, Indiana. Or at the one down in Maryland. We'd sometimes have to make a sample plan. If we failed what we'd submit, we'd make a corrective action either by disassembling all the units in that lot and make a corrective action and resubmitting another sample and retesting them for board phases, air-burst phases, function phases.

Here Bill's job as a witness took over, work which occurred in darkness: "And they only fired at night [so the] infrared could pick up the flash."[4.24]

Signifying a departure from his usual daylight schedule at Bristol, this nighttime shift occupied Bill's time into the 1960s when "we got thick and heavy" because of Vietnam. Ultimately, the increased regularity of the tests required the hiring of a full-time witness whom Bill personally trained "about the functions of the artillery fuse." Bill spent less time on the road for the remainder of his job at Ingraham. The expertise he developed on ordnance, however, served him well after his layoff; Timex in Waterbury eagerly employed him on its own fuse programs, guaranteeing him further work when Ingraham could not.[4.25]

Women and Class Consciousness at Ingraham

Despite their uniqueness, engineers in the early 1960s did not escape working-class status and still retained much in common with other laborers. In addition to sharing with them short-time work, layoffs, and wage rates that affected their actual earning potential, women blue-collar workers at Ingraham also underwent a reconceptualization of their socioeconomic identities, just as did the engineers. Women's position changed after 1945 as they developed their own ethic of breadwinning. In the highly competitive labor market of the postwar period, marked by a shrinking industrial sector and increasing desires for consumer goods and services, working women's struggle to make money and support a middle-class life against a perceived Soviet threat accented their roles as family wage earners. As a consequence, women felt increasingly entitled to job security and strong wages. They became more vocal in their efforts to achieve those aims, relying especially on their community roles as wives and mothers to justify such actions. "Helping," aiding one another in getting out production that characterized most all-women departments in the interwar years, diminished, at least where solidarities between single and married women were concerned. Walantyna Sakowski related that she and her coworkers "used to set their own quotas" during the 1940s but talked of a different scenario in later years. Sakowski remembered how as a pressworker she felt entitled to make as much money as possible to help her husband, a watchworker at the factory, to support their family. This meant the right to outperform other women on piecerates, especially if they were single. Sakowski said she reprimanded a slower woman, who could not compete with her: "'You know,' I says, 'You're single. You have yourself to take care of.' . . . I said, 'Remember I have a child. There's the doctor's bills. There's a dentist. There's always something.'"[4.26]

In an era that accentuated pronatalism, the family, and conspicuous consumption, Sakowski no longer felt the bonds of sisterhood so fervently. She saw herself in a different, more respectable, light when talking of household roles outside the rough, shop floor culture through which she asserted herself. Engineers, too, saw themselves as not simply workers, but as respectable professionals, even when not entirely so. In both instances, middle-class respectability predicated on the basis of wage earning imparted the changing notions of self. This common link suggests a new, if confused, awareness among Ingraham employees by the 1960s, one that as the example of engineers demonstrates blurred categories of working-class and middle-class distinction.[4.27]

At the same time, however, the demands placed upon women also led to an increasing willingness by female unionists to voice their own particular concerns during the 1960s, showing them to be champions of the working class. As before, women had been active supporters of their union, figuring prominently, for example, in the

Figure 4.11. Sessions' Pauline Yard, typical of the aging workforce, on the job, 1961.

winning of a closed shop in 1957. In a marked departure from recent trends, however, women orchestrated the only other strike during the decade after the 1960 strike to prevent Ingraham from relocating. Union men had previously dominated such public shows of class solidarity, as the 1960 strike symbolized. In late October 1966, however, women in the plating department staged a strike with far-reaching consequences for both their union and their factory. Thirty of them left their jobs in a wildcat strike upon learning that the firm planned to switch them from an hourly pay rate to piecework in violation of the current union contract. After having closed down the entire department, the women promoted their cause by leafleting other workers at the factory gates. Once their union, IUE Local 260, took up the grievance with management the women agreed to return to work pending the outcome of the negotiations.[4.28]

In the coming months, Local 260 and its president, Mrs. Mary Owens, acted decisively on the platers' behalf, especially when the company later discharged the women. In December the IUE filed charges against Ingraham and denounced the firm for dragging its feet during the arbitration after firing the women. "The Company," read a union statement, "in the opinion of the Local 260 Union's officers and the officers of the International Union, is at the present time engaged in unfair labor practices in its dealings with its employees and also shows complete disregard for the labor contract between the Company and its workers."[4.29]

Moreover, announced Mary Owens and her board, the local had also forced the company to abandon time-honored discriminatory employment practices that treated men and women's seniority separately because of pay differentials between them. In line with the Equal Pay Act of 1963 and the Economic Opportunity Act of 1964, Owens pursued this case before the Equal Economic Opportunity Commission (EEOC) beginning in 1965. The EEOC-mandated removal of different classifications for men and women meant that managers now had to revamp their payscales. The firm made a pact to give all women equal rights, with the company and the union agreeing to amend their contract to conform to the antidiscriminatory Title VII of the Civil Rights Act of 1964. Management immediately instituted job classifications whose pay rates reflected skill and effort, with all workers having equal opportunity to bid for job openings on the basis of seniority and ability. By pursuing the legal course they did, Mary Owens and other women unionists at Ingraham thus follow Dennis Deslippe's model of IUE women leading the charge for equal rights after the passage of the Civil Rights Act.[4.30]

New and Old Faces in the Workforce

The winning of job equality concluded Ingraham employees' wave of 1960s labor militancy. Thereafter, the workforce dealt with another change, which had its own effect on labor activism: the entry of nonwhites into its ranks. By late in the decade, employees at Ingraham revealed long ingrained notions of race when the company began to hire first a handful of African Americans and then Latino/as in response to the Civil Rights Movement.

While marginal at first, the hiring of nonwhites, namely Puerto Rican and African American, progressed steadily until the plant's closing under the name of Bussmann in 1991, according to superintendent George Power. At the factory's closing, "[w]e were higher than the quotas at the time," recalled George, estimating, "Oh, between Puerto Rican and black, 30 percent. . . . They started coming from New Britain [which had high Latino concentrations], a lot of them." The initial presence of these few men and women forced the traditionally white workforce to make explicit a deep-rooted belief that only Caucasians fulfilled the image of productive employees—a perception that many interviewees retained years after leaving Ingraham. Racial tensions never exploded due to the small numbers of nonwhites, but longtime workers nonetheless viewed life at Ingraham increasingly along the color line. Moreover, because of the marginal percentages of blacks and Hispanics, Caucasians dominated the bargaining table through the 1970s, literally whitewashing class activism.[4.31]

At Sessions Clock, by contrast, the story was one of a rapidly shrinking workforce that shielded employees from the larger society's shift on race. A workforce of roughly 100 made the hiring of new faces of color or otherwise highly improbable. In fact, as Sessions shrank, seniority and sex-typing of jobs worked together to make the workforce more exclusively older and female with considerable seniority (Table 4.1). In 1967, 62 percent of the women had worked at Sessions for at least 14 years and almost one in three had worked there for 19 years or more. By 1969 workers were on average 52 years old. Whereas homogeneity may have obscured race, however,

Table 4.1
Full Years of Service by Sessions Employees at Plant Closing, 1970

	Over 40	31-40	21-30	11-20	6-10	3-5	Under 3	TOTAL
Total	4	3	20	24	25	7	5	88
Women	2	2	16	22	22	6	5	75
Men	2	1	4	2	3	1	0	13

Source: Sessions Clock Company, "Full Years of Service," 1970, Beach, Calder, Anderson, and Alden Papers, Archives and Special Collections, University of Connecticut, Storrs, CT.

Table 4.2
Sessions Clock, Plant Averages by Union Negotiation, 1959-1967

	1959	1960	1961	1962	1963	1964	1965	1966	1967
Average	2.00	2.07	2.18	2.30	2.39	2.38	2.47	2.45	2.43

Source: Sessions Clock Company, computer-generated analysis of wages by plant, department, and all employees individually, 1959-1967, dated June 8, 1967, Beach, Calder, Anderson, and Alden Papers, 16.

the common age and gender among employees had a mobilizing effect in terms of labor activism. The majority were married or widowed, according to a 1960 sample. Veterans of childraising and family support, they knew the economic value of their paid labor. A central concern was achieving a living wage, which the women had incorporated into the agenda of Local 261 since its 1956 inception and enhanced by the establishment of a closed shop in 1958. Rather than getting a living wage, however, the union won small if incremental raises since that time (Table 4.2). As the final straw, a weak Sessions management could not afford wage increases for all workers under a wage reopener in the union contract that occurred in 1967. While the union sought a 20-cent-an-hour increase, management proposed at first only a raise for men and in the end only a pitiful 3 cents more for everyone. Against the "insult," as union secretary Alma Charpentier deemed the final offer, women took to the picket lines to defend themselves with the same conviction that their sisters at Ingraham's had shown in their labor activism.[4.32]

Women's Militancy at Sessions Clock, May-June 1967

At 7 a.m. on May 5, 1967, the 99 women and 11 men of IUE Local 261 walked off their jobs at Sessions. "LOCAL 261 ON STRIKE FOR A DECENT WAGE" screamed one of the most conspicuous picket signs on the first day of the walkout. The strike ended nine weeks later after violence, injuries, and several arrests unparalleled in local history. What sustained the union's resolve was clearly members' sense of mutual obligation. Years of working together with neighbors and, in many cases, family members created a sense of solidarity on the job, at home, and in the outside community. At the same time, the changing face of management from family owner/managers to impersonal, outside professionals facilitated the workers' stand. Unlike the locally based Sessions family, the new owner, Consolidated Electronics, was a conglomerate of over 60 companies and depersonalized management; personnel at the highest levels of management regularly arrived from outside Bristol. As the 1960s progressed, outsiders, working as disinterested professionals, undermined the close-knit character of earlier times, and opened the door to worker unrest. Under these circumstances, the women, already concerned about earning a living wage, could justify striking despite years of loyal service—or even because of it if they felt that they had earned better treatment, which many did.[4.33]

Because they derived their motivation from their place in the paid labor force, the women did not see themselves as radicals. The intense social questioning of the late 1960s did not strike a chord in Forestville and Bristol in 1967, especially among the older individuals at Sessions who focused their attention on familial survival.[4.34] Rather, the workers believed that they behaved well within traditional bounds by agitating for higher wages as wives, mothers, and widows. So strong was this idea that 30 years later former union president Kitty Baldaccini still did not see how the strikers' activities could be construed as unwomanly. Beginning on May 12, the company called on local police three times to escort delivery trucks driven by management personnel through the picket line in order to keep the factory functioning.

Figure 4.12. IUE Local 261 at the start of the 1967 strike. *Bristol Press*, May 5, 1967.

On the third occasion, violence erupted between workers and authorities. In remembering the clash, Kitty could not understand why management ever needed police. "We're women," she reasoned. "And all the women are over 50 years old." Mockingly, she added, "So I'm sure we could manhandle [management]." Kitty's comment was ironic, casting aspersions on management's manhood, but it also conveyed her belief that the women were not threats. In calling the police, managers thought otherwise. As soon as the women lent their standing as family members to activity perceived as behaving outside the norm, management treated them as menaces to their interests.[4.35]

Although the women remained verbally combative throughout the strike, the police had previously refused to take physical action against them. Instead, the police acted only when provoked by a man, in this case IUE field representative James Ingalls, who was in Forestville overseeing the strike at Sessions as he had the 1956 founding of Local 261. When police attempted to disperse the crowd of 40 or 50 pickets barricading Sessions' entrance, Ingalls reportedly lunged forward and declared, "Don't you dare hit those women." With this remark, police threw Ingalls to the ground and beat him so badly that he required hospitalization. Whereas the women's femininity had protected them, Ingalls' masculinity put him at odds with male officers. Police met Ingalls' action to protect women as an affront to their own manliness, one which had to be met with a "male" response of violence.[4.36]

The police assault on Ingalls touched off a general melee. In the ensuing fracas, one policeman suffered a fractured right hand, reportedly from Ingalls fighting back, while at least three women received bad bruises from police; one "had all black and blues on her arm," as Kitty remembered. The police also made several arrests for breach of the peace: Ingalls, his wife Pauline who attacked officers beating her husband, and Clayton Aiudi, the husband of a striker who happened to be at Sessions in support of the women.[4.37]

As Aiudi's presence suggests, family and friends could sustain the pickets but they could also divide ranks. Floor worker Lois Cieszynski had a brother, John Mastrianni, who worked in upper-middle management. Prior to the strike, Lois' coworkers jokingly referred to her as a company pet because of John's position. During the strike, this married mother of a small boy drew the attention of other strikers despite her participation because her brother was a negotiator for the firm. Unionists accused Lois of informing management of the strikers' strategies, especially since the bosses met at her brother's house, and forced her to picket in front of his office. Lois could not forgive the indignation, which she remembered as "so stupid." She blamed Kitty and confronted her openly. "I threw her against the wall down at John's office," Lois recalled. "I would have beat the living daylights out of her." John, who had his tires slashed during the strike, prevented a physical battle by telling her to try to get off the picket line, which she did after securing a doctor's excuse. In siding with her brother, Lois withdrew from the ranks of women alongside whom she had worked for almost 20 years. Her example shows that family and neighborhood could undermine as well as support solidarity based on gender and class.[4.38]

Despite division on the picket line, the local maintained a unified façade during the strike, especially when handed an injunction which management asked for on May 25 in the wake of violence and mass picketing that included Ingraham's Local 260. Despite the setback, members still maintained outward unity against management during the three meetings that highlighted the remainder of the strike. Held at the Connecticut State Labor Department in Wethersfield and attended by state and federal mediators, these meetings concluded on July 7 when the union won a 10-cent-an-hour increase in wages retroactive to February 15, 1967, when the old wage clause expired. Worker solidarity won more than triple management's final prestrike offer of 3 cents.[4.39]

Sessions' Demise and Union Scapegoating

After the conclusion of the strike vote on May 3, Sessions' assistant general manager Harry Miller promised that should the women go forward with their planned walkout, the company would close down within a year. Because of the firm's already ailing condition, Miller predicted that any walkout, especially if prolonged, would deal a fatal blow to the plant's remaining profitability. On the surface, the strike indeed sounded the death knell for Sessions. In May 1968 Consolidated Electronics sold all clock production assets in the factory to the United Clock Company in Brooklyn, NY, and stopped timepiece manufacturing in October when employment stood at 140. The remainder of the plant concentrated on making timers until Consolidated Electronics merged Sessions with North American Phillips Corporation (Norelco) of Wilmington in August 1969 and moved its operations to one of its subsidiaries in Frederick, MD. The plant closed finally on July 1, 1970, putting out of work the remaining 75 employees. Sessions had received no orders for electronics products for months and had incurred deep operating losses since the end of 1969. Publicly, the corporation focused on the economic reasons for dismantling Sessions. Privately, and in all interviews conducted with both former workers and managers associated with the strike, the corporation cited worker militancy as a primary reason.[4.40]

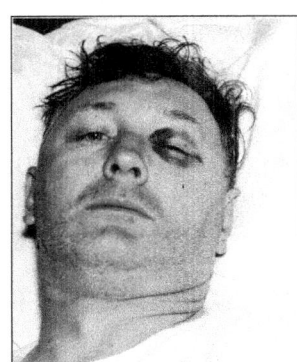

Figure 4.13. James Ingalls after being beaten by police, May 19, 1967. JAMES INGALLS PAPERS, ARCHIVES AND SPECIAL COLLECTIONS, UNIVERSITY OF CONNECTICUT AT STORRS.

Because of the consensus that the strike closed the factory, and because they now faced losing their jobs, workers looked for a scapegoat. The workforce found its target in arguably the most visible activist of the ordeal, union president Kitty Baldaccini. Long before the strike, Kitty had proven herself a president not afraid of speaking her mind. In her interview, she said she challenged anybody whom she thought was wrong—even President John Kennedy with whom she publicly took issue at a conference on tariffs, a subject of which she, as union president, had some knowledge. The same boldness led Kitty to clash regularly with the assistant general manager, whom she viewed as an unfair boss well before the strike. "And things we thought were pretty good until we got this Mr. Miller," Kitty recounted. "He was a son of a gun. Of course, I was president of the union. And everything was going good. They liked me. . . . it was because I was honest, fair and square. And, of course, I never lost a grievance. But when this Miller came in, well, then it was a different thing. I think he was really trying to break the union because he was giving me a rough time sometimes."[4.41]

Kitty disliked Miller and cited his antagonistic, union-busting attitude as the reason for the strike. "And [our contract] was supposed to be for two years but he [Miller] says that they couldn't afford to give us a raise," she explained. Miller offered an olive branch, saying "'But we'll open the contract just for wages next year.'" As president, she took his words in good faith, only to be rebuked:

And we [the union] says, 'Okay.' Well, when the time came, he refused. He wouldn't even consider it. And so, that's when we went on strike. And it's a shame because even if he had offered us a nickel, it would have been alright because he made an agreement. And if you don't honor your agreement, what good is a union? And so, that's when we went on strike. And they were very nasty. And they got the police in and they really mauled a few of them.

Kitty felt that only by striking the union retained its dignity.[4.42]

Others felt differently. Years later, Lois Cieszynski claimed that Kitty alone bore responsibility for the event, declaring "Well, Kitty is the one that wanted a strike." Laura Santago, Lois's sister, felt the same way. "Up to this day, see, I have no use for Kitty," she said unabashedly. "I really don't and you can go back and tell her I said it, too, [because] she was NG [no good]." To label Kitty as the source of the strike and the cause of Sessions' demise of course misrepresents and inappropriately personalizes the activism of the women who agreed to the walkout. Indeed, the tense relationship between the union president and Miller did not help matters. Management interviewees confirm a fierce animosity between the two. Frederick Shores, who worked two stints between 1961 and 1968, first in production and then as Miller's protégé in sales, and who was also a member of the company negotiating team, best summarized the strength of feeling that existed between the two: "And I guess it got to be a situation of whatever you want, you're not going to get. And whatever you give me, I don't really want; I want something else." Frederick also validated Kitty's claim that Miller refused to accommodate the union in any way during the strike; he remembered a management official explaining, "'You can settle this strike for a nickel.' And Harry Miller, Harry said no." Shores had harsh words for his mentor's intransigence: "A nickel. Seventeen weeks. You know, brutality, people getting beat up and God knows what else for a nickel." Scapegoating Kitty distorted the historical record, creating the false image of a rank-and-file as passive victims of the walkout blindly following their president and ignoring the common purpose and solidarity that allowed the women of Local 261 to strike in the first place.[4.43]

As layoffs began, workers had a more pressing concern: finding new jobs. Former strikers, however, received little welcome from the outside community. Kitty quickly realized that she could not get a job from local employers. An initial position at Ingraham ended shortly after a union officer there recognized her. Apparently seeing her as a troublemaker, he informed management of her role as union president during the strike. Kitty finally got steady employment at toolmaker Stanley Works in New Britain where nobody recognized her. However, she achieved job security at a price; she explained that at the time she so worried about her identity that she denied that she had ever been in a union before and paid 35 dollars to join Stanley's local. Less visible strikers also felt the power of the blacklist. Lois tried to get a job at such places as GE Plainville, the local General Electric plant, and Bristol's Superior Electric. She was unsuccessful there and elsewhere. "I couldn't even get hired in a meat store," Lois lamented. "I wanted to go to Stop and Shop." After she reported these places to the local employment office in Meriden, all offered her jobs. Lois said she deliberately took GE because "they refused to hire me under any circumstances and I went there every other day." She was still at GE when interviewed in 1997.[4.44]

Conclusion

Kitty's and Lois' experiences reflect the wounds left by the Sessions strike in 1967. Not only could neither find a job but Kitty also lost one through betrayal by a fellow unionist. These conditions constituted the cost of striking violently. The changing cultural perceptions of women allowed those in the Bristol area a certain latitude in their postwar pursuit of breadwinner wages. The increased activism of women in Ingraham's Local 260, for example, met little public opposition. Presumably, the women's ability to achieve greater gender equality without the prolonged striking and violence that their sisters at Sessions endured put their activities within the realm of acceptable norms. The profundity of the strike staged at Sessions, however, was so intense that it still resonated 30 years later when interviewees displayed scars of the event and vehemently took sides. Its memory still stirred lively debate in the individuals whose lives it affected, long after Sessions Clock closed and even to the exclusion of the issue of race that shaped worker identity at Ingraham in its final years.

Because it involved women, the strike also demonstrates the tendency to blame gender for the deindustrialization that ultimately claimed Sessions. Already in the works in Bristol's clock and watch industry, deindustrialization would have closed Sessions whether the women struck or not. Given the character of the strike, however, it was easier for some to blame women than face the reality of the situation. For this reason, the memory of the Sessions strike is both a lesson about how gender biases can distort the historical record and a reminder of the need to constantly challenge these inequalities in order to provide a fairer telling of history in the future.

Epilogue

Time Runs Out: The End of Work

The closing of Sessions Clock Company in 1970 signaled the end of timepiece manufacturing in "Clock City," although Ingraham existed in name until its 1974 reorganization. While some scapegoated the union for the demise of the local industry, many interviewees saw what happened in Bristol as a microcosm of the larger trends toward a postindustrial world. "Because when actually they moved," Kitty Baldaccini reminisced of Sessions, "they got most of the movements from overseas. So, all we did was put them together." Kitty had seen the writing on the wall since the early 1960s when, as union president, she had lobbied President Kennedy for higher tariffs at a labor conference. She found that her protests resonated with a fellow IUE member from the rubber-producing center of nearby Naugatuck, whom she met about six months later. She recalled:

> And he says, 'You don't remember me,' he says, 'but I was at that seminar.' And he says, 'You were right.' He said he worked in the rubber factory in Naugatuck. He says, 'It's all gone now.' I says, 'Because it stands to reason if you're going to get something from overseas and you pay a dollar for it and if it's made here it will cost you two dollars, they go for that and it's not helping industry here in the United States, you know?'

Kitty understood the meaning of imports for Sessions. "But that hurt us an awful lot because everything was imported," she continued. "That's why they couldn't make the motors anymore. Because they figured it was cheaper to import them. And, of course, our place was real big at one time and it just dwindled down."[5.1]

George Power, the former Ingraham IUE activist and later production superintendent, offered similar insights for his factory. George linked Ingraham's decline to the inability to reconcile high wages with the changing times. "Clocks-and-watches wages were the first thing that affected clocks and watches because it took so many man hours to put a clock together," he explained. He realized that "when you competed for wages, foreign imports would kill clocks and watches first because with the hours it took just to make" the "wages would become a big issue, especially when added to the cost of materials. So there was only so much you could cut down on material and all that stuff." George saw the special relevance for Ingraham as a producer of military ordnance. "This is why when we made the military fuses they were talking about putting a tariff on and all that," he remembered, "so they wouldn't wipe out the clock and watch work because they knew that business was going to get wiped out from imports because you just couldn't compete with that." Burdened by imports, George concluded, observers began asking, "If you lose the Ingrahams and then all these other clocks, who's going to make the military fuses?"[5.2]

By 1970 the federal government, long passive on the issue, numbered among the concerned and took matters into its own hands. The worried Department of Defense set its sights on the failing Hamilton Watch Company in Lancaster, PA, which, along with Ingraham Industries and the famed Bulova watchmakers, was the remaining supplier of the essential ordnance. In 1971 the government purchased the precision watchmaking facilities at Hamilton to guarantee a future supply of fuses. The move saved Hamilton, and the jobs of its workforce, but it placed the firm's mechanical time fuses and other time devices on a noncompetitive basis, ruling out a chance for army contracts from either Bulova or Ingraham. Octogenarian Dudley Ingraham protested the purchase and had the help of Connecticut's U.S. Representative (and future governor) Ella Grasso, who openly opposed the sale in Washington. Grasso's pleas, however, were trumped by worries about national security amid the Cold War, and local concerns yielded to military interests. The gov-

Figure E.1. Sessions Clock Factory, Forestville, CT, summer 2004.

ernment's purchase boded ill for what remained of Ingraham, as the outcomes of 1974 proved.[5.3]

In retrospect, the loss of the local industry and the jobs it provided could have been prevented with foresight. The old owners, the Ingrahams and Sessionses, would have likely called for free trade amid higher tariffs as the solution. Yet the twentieth-century history of their industry and its demise shows that free trade was not the answer. Rather, the lessons of Bristol's clock and watch industry called for a state industrial policy to actively defend the interests of working Americans, who were the very backbone of manufacturing. Such a proactive measure at the federal level could have better ensured job protection as well as a longer life for one of Connecticut's defining industries.

Figure E.2. Dudley Ingraham by William Draper, 1965.

Figure E.3. Kay Laviero (left) and Dottie Beaucar (right), Christmas 2002.

In the absence of such a policy, local workers developed strategies and took new jobs to continue to support themselves and their families. Memories of what once was, however, remained. For older, well-seasoned individuals, reconciling recollections of past high wages and steady employment with the increasingly service-oriented, unstable economy of the present proved hard. Former adjusters, for example, had spent long working lives at the top echelons of pay. Now, they struggled alongside younger workers to keep jobs that promised neither long-term security nor high pay. Some tried defensively to resist the transformation.

Bill Casey from Ingraham was one of these. George Power preserved Casey's story for posterity, framing it within the context of Ingraham's 1964 downsizing to the smaller Redstone Hill plant, a circumstance itself indicative of the local industry's ebb. "So we went down to the new [factory on] Redstone Hill Road," Jack rhapsodized:

> I'll always look back memorably, we had Bill Casey. He was our last pocket watch adjuster. I finally said to him, we finished our last clocks and I said, 'Bill, come Monday, I'll give you something else,' and he says, 'I'm an adjuster.' We used to have hundreds and hundreds of adjusters, and once you got to be an adjuster, you're king. Make good money and all that. I says, 'There's no more adjusting.' I mean, we didn't have nothing. He had been there, it was maybe fifty years already.

> Neither George nor Bill's supervisor could get him to understand that the times had changed: He was the last of the old-timers and he was sitting at his bench and the supervisor came over to me and says, 'I tried to get Bill and give him something else to do, and he says he's an adjuster. I said leave him alone.' He was just at a stage of his life that he couldn't understand after having [so many] adjusters, there was no more adjusting to be done.

George gave "him little things to do" to keep the old man busy, especially since recent years had not been kind to Bill. "He had lost his wife, you see," George explained. "At . . . that time of his life, everything was going and it was a tough thing."[5.4]

Everything was, of course, going, not just for Bill but for the industry he represented as well. "[So] I let him sit there," George continued. "He'd come in and he'd just sit at his bench and poke around and all that, and then it sunk in and I had some little, little things I'd bring over to him to do and all that because he was the last of an old breed and he just couldn't accept it and basically let him retire."[5.5]

Bill did not retire alone. With him went the remnants of Bristol's clock and watch industry, a once vibrant force in the city's economy and social life just like the adjusters it employed but now old and feeble like Bill himself. The memory of life at Ingraham and Sessions, however, did not share the same fate. It continues to run deep in the Bristol area, preserved by the memories of the women and men who ordered lives and identities around employment within those factories. The local industry left its imprint on so many workers, but so, too, did the same people leave their own individual and collective marks on Ingraham and Sessions Clock. Those impressions gave life to the factories and forged memories that invariably withstood the test of time. Time did not erase those human links to the past.

Notes

Introduction

I.1. Sources that span the growth and decline of Bristol's clock and watch industry from 1820 through the twentieth century include Bruce Clouette and Mathew Roth, *Bristol, Connecticut: A Bicentennial History, 1785-1985* (Canaan, NH: Phoenix Publishing, 1984); Carleton Beals, *The Making of Bristol: Our Yankee Heritage* (Bristol, CT: Bristol Public Library, 1954) and Epaphroditus Peck, *A History of Bristol, Connecticut* (Hartford: The Lewis Street Bookshop, 1932).

I.2. Interview by author with Lillian Rock, October 23, 1996. For examples of gendered work cultures elsewhere, consult Lisa M. Fine, *The Story of Reo Joe: Work, Kin, and Community in Autotown, U.S.A.* (Philadelphia: Temple University Press, 2004); Ardis Cameron, *Radicals of the Worst Sort: Laboring Women in Lawrence, Massachusetts, 1860-1912* (Urbana: University of Illinois Press, 1993); Joy Parr, *The Gender of Breadwinners: Women, Men, and Change in Two Industrial Towns* (Toronto: University of Toronto Press, 1990); and Susan Porter Benson, *Counter Cultures: Saleswomen, Managers, and Customers in American Department Stores, 1890-1940* (Urbana: University of Illinois Pres, 1987). Linda Nicholson's edited *Feminism/Postmodernism* (New York: Routledge, 1990) provides an introduction to the concept of gender as a constantly constituted and reconstituted aspect of people's lived experiences.

I.3. Examples of this literature include Tran Duy Ly, *Ingraham Clocks and Watches* (Fairfax, VA: Arlington Books, 1998) and *Sessions Clocks* (Fairfax, VA: Arlington Books, 2001); Snowden Taylor and Kenneth D. Roberts, *Forestville Clockmakers* (Fitzwilliam, NH: Ken Roberts Publishing Company, 1992); Donald R. Hoke's chapter on wooden movements in Connecticut in *Ingenious Yankees: The Rise of the American System of Manufactures in the Private Sector* (New York: Columbia University Press, 1990); John Joseph Murphy, "Establishment of the American Clock Industry: A Study in Entrepreneurial History" (unpublished Ph.D. dissertation, Yale University, 1961); Barrows Mussey, *Young Father Time: A Yankee Portrait* (New York: Newcomen Society in North America, 1950); Edward Ingraham, "Clockmaking in Connecticut," *The Antiquarian: The Bulletin of the Antiquarian and Landmarks Society, Inc., of Connecticut*, Vol. III (1) (June 1951): 10-16; and Lockwood Barr, "Clockmakers of Bristol" (unpublished manuscript in E. Ingraham Company Papers at the University of Connecticut).

I.4. On the relationship of timepieces to capitalism, see Martin Breugel, "'Time That Can Be Relied Upon:' The Evolution of Time Consciousness in the Mid-Hudson Valley, 1780-1860," *Journal of Social History* (Spring 1995): 547-564, and more recent work by Alexis McCrossen. In seeking to understand the evolution of time consciousness, Breugel argues that clocks and watches served to organize the processes of production in a capitalistic society whose division of labor was growing increasingly complex. McCrossen updates Breugel's inquiry in "The 'Very Delicate Construction' of Pocket Watches and Time Consciousness in the Nineteenth Century United States," *Winterthur Portfolio* 44 (Spring 2010) 1, by offering a more national look at how pocket watches advanced existing temporal sensibilities through increased circulation after the 1830s. McCrossen's *Marking Modern Times: A History of Clocks, Watches, and Other Timekeepers in American Life* (University of Chicago Press, 2013) considers public clocks and their role amid nineteenth century capitalism in habituating U. S. citizens to "clock time," or the practices and material culture related to mechanical timepieces. In examining a broad range of examples that include bells, mechanical clocks, and time balls, McCrossen draws upon the theoretical and historical research of recent anthropologists and historians to showcase the evolution of "modern time discipline," defined in her analysis as a cultural system whereby clocks become the nexus for organizing and controlling society's temporal awareness.

I.5. Roy Rosenzweig, *Eight Hours for What We Will: Workers and Leisure in an Industrial City, 1870-1920* (Cambridge, UK: Cambridge University Press, 1983).

I.6. See David Landes, *Revolution in Time: Clocks and the Making of the Modern World* (Cambridge, MA: Harvard University Press, Belknap Press, 1983). Landes updated this classic work with a revised and enlarged edition, published by Belknap in 2000. This edition includes more recent findings about medieval and early-modern timekeeping, as well as contemporary high-tech uses of the watch.

I.7. Chris H. Bailey, *Timepiece Journal*, No. 5 (Winter 1996): 8, frontpiece editorial.

I.8. Helpful in understanding the value of oral history is the theoretical literature on the subject. See, for instance, David King Dunaway and Willa K. Baum, editors, *Oral History: An Interdisciplinary Anthology* (Walnut Creek, CA: AltaMira Press, 1996); Sherna Berger Gluck and Daphne Patai, editors, *Women's Words: The Feminist Practice of Oral History* (New York: Routledge, 1991); and Alessandro Portelli, *The Death of Luigi Trastulli and Other Stories: Form and Meaning in Oral History* (Albany: State University of New York Press, 1990).

Chapter 1

1.1. For examinations of paternalism at other companies, see Gerald Zahavi, *Workers, Managers, and Wel-*

fare Capitalism: The Shoeworkers and Tanners of Endicott Johnson, 1890-1950* (Urbana: University of Illinois Press, 1988); Joseph M. Turrini, "The Newton Steel Strike: A Watershed in the CIO's Failure to Organize 'Little Steel,'" *Labor History*, Vol. 38 (2-3) (Spring/Summer 1997): 229-265; Mary Lethert Wingerd, "Rethinking Paternalism: Power and Parochialism in a Southern Mill Village," *Journal of American History*, No. 83 (December 1996): 872-902; and David Montgomery et al.'s discussion of employer paternalism and welfare work in "Patronage, Paternalism, and Company Welfare," *International Labor and Working Class History Journal*, No. 53 (Spring 1998).

1.2. John Joseph Murphy, "Establishment of the American Clock Industry: A Study in Entrepreneurial History" (unpublished Ph.D. dissertation, Yale University, 1961); 1940 factory footage of E. Ingraham Company, Edward Ingraham Library, American Clock and Watch Museum, Bristol, CT.

1.3. Bruce Clouette and Mathew Roth, *Bristol, Connecticut: A Bicentennial History, 1785-1985* (Canaan, NH: Phoenix Publishing, 1984): 106, 91-93, 95; Murphy, "Establishment of the American Clock Industry," Chapter VII.

1.4. Census returns for 1880, 1910, and 1920.

1.5. "Schedule 3, Manufacturers," manuscript of census returns for Bristol, CT, Tenth U.S. Census, 1880; office memo, December 18, 1920, Dudley S. Ingraham Papers, Series I, Box 1, Archives and Special Collections, University of Connecticut, Storrs, CT—hereafter referred to as "Dudley Ingraham Papers"; summary of labor turnover rates among Bristol employers for April 1928, Bristol, May 1, 1928, Dudley Ingraham Papers, III, 33; office memo, March 26, 1928; labor turnover reports, August 1922-April 1928; E. Ingraham Company—hereafter referred to as "Ingraham Company," "Age Groups in the E. Ingraham Company Employ March 7, 1940;" all E. Ingraham Company Papers, Archives and Special Collections, University of Connecticut, Storrs—hereafter referred to as "Ingraham Company Papers." For examples of Ingraham timepieces, consult Tran Duy Ly's *Ingraham Clocks and Watches* (Fairfax, VA: Arlington Books, 1998).

1.6. Photocopy of letter by Elias Ingraham to "Irvan My Dear Grand Son," December 25, 1884, in Ingraham family bible, author's family.

1.7. *Bristol Herald*, June 19, 1897; Carleton Beals, *The Making of Bristol: Our Yankee Heritage* (Bristol, CT: Bristol Public Library, 1954): 210; Ingraham Company, *The Ingraham Story* (Bristol, CT: The E. Ingraham Company, circa 1954): 10, 20; Ingraham Company, "Total Watches Delivered Per Year: Pocket, Wrist, Ladies, Delivered from the Watch Department to the Shipping Room," Ingraham Company Papers, Box 298; *Bristol Press*, June 3, 1940; Ingraham Company, All Available Statistics on Clock and Watch Industry," circa 1946, author's collection. The term "clock watches" is generally understood to mean inexpensive watches made in a clock factory, practically all being made from "stampings," and having no jewels—like Ingraham's.

1.8. Howard E. Kneeland, "The History of the Sessions Clock Company, 1829-1966" (typescript manuscript, courtesy of Francis Savage); Chris H. Bailey, "The End of an Era: The Sessions Clock Company," in the reprinted *Sessions Clocks Catalogue No. 65, 1915* (Bristol, CT: American Clock and Watch Museum, Inc., 1977).

1.9. Howard E. Kneeland, "The History of the Sessions Clock Company"; Chris H. Bailey, "The End of an Era," *Bristol Press*, May 3, November 4, 1930. Tran Duy Ly provides a comprehensive listing of Sessions' clocks in *Sessions Clocks* (Fairfax, VA: Arlington Books, 2001).

1.10. "Clocks, Watches, and Materials and Parts (Except Watchcases)," *Sixteenth Census of the United States; 1940: Manufactures: 1939, Vol. II, Part 2* (Washington, DC, 1942): 317-319, 317. In both 1909 and 1919, the United States listed 91 firms, but the number did temporarily drop to 88 in 1914.

1.11. Drawn from figures for 1930—the first full year of the Depression, initial unemployment in clock and watch factories nationwide amounted to 1,018 men and 528 women. Source: "Table 3: Unemployment Classes A and B, by Sex and Occupation, for the United States: 1930," *Fifteenth Census of the United States: 1930, Unemployment, Volume II, General Report* (Washington, DC, 1933): 15.

1.12. Edward Ingraham, "Reminiscences of Edward Ingraham, Clockmaker," Paper IV, 14, Edward Ingraham Papers, Series IV, Box 24; Archives and Special Collections, University of Connecticut, Storrs—hereafter referred to as "Edward Ingraham Papers."

1.13. *Bristol Press*, May 10, 1941; Bailey, "Sessions' Employee Numbers, 1930-1953," *Sessions Clocks Catalogue No. 65;* summary of labor turnover rates for April 1928, in Dudley Ingraham Papers, III, 33.

1.14. *Bristol, Connecticut in World War II . . . depicting the part played by our brave men and women who dedicated their services to their country, as well as the mobilization of our home life and industries in support of the war effort, 1939-1946* (Worcester: [Bristol] World War II Historical Committee, 1947): 328, 312, 324; Beals, *Making of Bristol*, 260; *The Ingraham Story*, 14, 11; *Bristol Press*, June 3, 1940; August 14, 1941; May 22, June 16, and November 28, 1944.

1.15. "Automatic Screw Machines in Use as of July 22, 1944," July 26, 1944, Ingraham Company Papers.

1.16. On tariff and tax legislation in this context, together with the effects of World War II on Ingraham and Sessions, see Philip Samponaro, "Taxes, Tariffs, and Timepieces: International Trade and the Clock and Watch Industry in Bristol, CT, 1936-1955," *Connecticut History*, Vol. 48 (1) (Spring 2009): 1-29. See also David Landes' examination of domestic tariff policy and the American clock and watch industry from the 1930s onward, in *Revolution in Time: Clocks and the Making of the Modern World* (Cambridge, MA: Harvard University Press/Belknap Press, 1983).

1.17. Copy of circular issued by German agents, in Walter Camp to W. S. Ingraham, March 17, 1921, Ingraham

Company Papers, II, 11; Public—No. 316—73D Congress [H. R. 8687], "An Act to Amend the Tariff Act of 1930;" transcript and sound recording of Dudley S. Ingraham on "Paths to Prosperity," broadcast on NBC's Blue Network, March 13, 1938, 7 p.m. EST, author's collection; Edward Ingraham, "Reminiscences of Edward Ingraham, Clockmaker," Paper IV, 15, Edward Ingraham Papers, IV, 3; Ingraham Company, undated memo, circa 1938, Ingraham Company Papers.

1.18. "The Trade Agreements and the Domestic Watch Industry," circa 1945, Ingraham Company Papers; Ingraham Company, "All Available Statistics on Clock and Watch Industry," author's collection; Chris H. Bailey, "Sessions Clock Company's Annual Production, 1930-1953," *Sessions Clocks Catalogue No. 65,* 117.

1.19. Dudley Ingraham to Messrs. Edward M. Greene, Edward T. Carmody, November 21, 1951, Edward Ingraham Papers, I, 9; Dudley Ingraham to author's parents, 1977, author's collection.

1.20. Ingraham Company memo, "Excise Tax," Ingraham Company Papers, III, 12; Bertrand Hull to Dudley Ingraham Papers, January 30, 1945, Dudley Ingraham Papers; *Bristol Press*, July 28, 29, 31, 1954. Theodore Kovaleff treats the involvement of the Eisenhower administration with domestic industry in *Business and Government during the Eisenhower Administration: A Study of the Anti-Trust Division of the Justice Department* (Athens: Ohio State University Press, 1980).

1.21. W. K. Sessions, "The Sessions Clock Company," Exhibit A, in Reconstruction Finance Corporation Form L-143, "Information Required as to Each Borrower," Record Group 234, Reconstruction Finance Corporation Examining Division, Paid Loan Case Files, 1932-1942 (PI 173/ Entry 85): American Community Financial Corporation: NRA Mortgage Loan #11: Sessions Clock Company, Forestville, Connecticut, 570, 67:9:6, Box 22, National Archives, Washington, DC; *Bristol Press,* August 15, 1922, August 25, 1923, September 11, 12, 1924; interviews by author with Laura Santago, Sandy Giammateo, and Domenic Giammateo, April 9, 1997; Lois Cieszynski, March 18, 1997.

1.22. *Bristol Press,* August 27, 1920, February 14, 1921, April 13, 1876, January 17, 1878, May 23, 1893, January 5, 1905, November 1, 1906, November 1, 1910; May 14, 1930; Beals, *Making of Bristol*, 193.

1.23. *Bristol Press,* March 24, 1904, January 17, 1910; Beals, *Making of Bristol,* 226.

1.24. *Bristol Press,* April 27, 1926; Elias to Irving Ingraham, Christmas, 1884.

1.25. Interview with anonymous woman, "Bertha D.," by author, March 17, 1997.

1.26. *Bristol Press,* July 3, 1939, May 23, 1945, June 28, 1939. For other regional and state boards on which Edward served, see *Bristol Press,* October 31, 1946; January 22, July 22, October 14, and November 19, 1948. On Dudley, consult *Bristol Press,* February 4, 1941. Chris H. Bailey tells the story of the American Clock and Watch Museum in *Fifty Years of Time: The First 50 Years of the American Clock and Watch Museum* (Bristol, CT: American Clock and Watch Museum, 2009).

1.27. *Bristol Press*, December 10, 1946; December 10, 1947; January 17, 1949; Manufacturers Association of Connecticut, *Connecticut Industry* (1947-1948). As member companies, the Clock Manufacturers Association included the General Electric Company of Ashland, MA, the William L. Gilbert Clock Corporation of Winsted, CT, the Herschede Hall Clock Company of Cincinnati, Ingraham, the Lux Clock Manufacturing Company of Waterbury, CT, the New Haven Clock Company, Sessions Clock, Seth Thomas Division of General Time Instruments Corporation of Thomaston, CT, Warren Telechron Company of Ashland, MA, and the Westclox Division of General Time Instruments Corporation in LaSalle, IL. Collectively, the companies manufactured approximately 90 percent of all clocks and all "clock watches" produced in America by 1941. Although the term "clock watches" usually denoted nonjeweled movements, roughly 10 percent of those produced by members of the Clock Manufacturers Association did have jewels as of 1941.

1.28. Edward Ingraham, "A Personal Glimpse into the E. Ingraham Company and Clock Manufacturing in the 20th Century," *Timepiece Journal,* No. 5 (8) (Winter 1996): 151-162, 154; interview with Robert Tetro by author, October 30, 1996; *Bristol Press*, October 16, 1929; July 14, December 15, 1936.

1.29. *Bristol Press*, June 3, 1940.

1.30. Clock Manufacturers Association of America, typescript, September 24, 1941, Edward Ingraham Papers, V, 26; Ingraham Company, office memo, March 27, 1928, Ingraham Company Papers, III, 12.

1.31. U.S. government statistic on standard of living as quoted in Ingraham Union News, circa March 26-April 30, 1941; Ingraham Company, office memo, "% Change in Hourly Rate 1939-1940;" "The E. Ingraham Company, Analysis of Employee's Earnings," August 10, 1942, all Ingraham Company Papers, VI, 290. On the cost of living in the 1930s, consult Thomas A. Stapleford, "'Housewife vs. Economist': Gender, Class, and Domestic Economic Knowledge in Twentieth-Century America," *Labor: Studies in Working-Class History of the Americas,* No. 1 (2) (Summer 2004): 89-118.

1.32. Edward Ingraham, "Four Generations of Ingraham Management, 1805-1885," 19, Edward Ingraham Papers, IV, 22; Beals, *Making of Bristol*, 260; Tetro interview; *Bristol Press,* January 5, 1929; Dudley S. Ingraham, "Paths to Prosperity through Employment," March 13, 1938, Dudley Ingraham Papers.

1.33. Antonio Gramsci, "Americanism and Fordism" in *Selections from the Prison Notebooks* (New York: International Publishers Company, 1971), edited by Quinton Hoare and Geoffery Nowell Smith.

1.34. Santago, Giammateo, and Giammateo interview;

interviews by author with Catherine "Kitty" Baladaccini and Hazel Mongillo, June 4, 1997; Kathryn Laviero and Mary Budnik, October 16, 1996.

1.35. Interview with anonymous woman, "Bertha D."; Dudley Ingraham to Mr. H. H. Crimmel, August 3, 1933, author's collection; Edward Ingraham, "Reminiscences of Edward Ingraham, Clockmaker," Paper IV, 13-14, Edward Ingraham Papers, IV, 3; Exhibit A of W. K. Sessions, "The Sessions Clock Company"; Edward Ingraham to Dudley Ingraham, March 4, 1938, Edward Ingraham Papers, I, 1; *Bristol Press*, September 18, 19, 24, October 30, 1934.

1.36. Zahavi, *Workers, Managers, and Welfare Capitalism*, 99-119.

1.37. U.S. Census, *Composition and Characteristics of the Population, 1910* (Washington, DC, 1913): 590; Clouette and Roth, Bristol, Connecticut, 140-141. Another 7.3 percent of Bristol's population in 1910 was "mixed," meaning persons with one native-born and one immigrant parent or with two immigrant parents from different countries.

1.38. Interview with Kenneth Webster by author, October 30, 1996.

1.39. Edward Ingraham, "Notes prepared for use in oral argument at hearing in executive session of the Labor Committee of the [Connecticut] State legislature considering bill for a shorter work week for women . . . ," May 2, 1935, Dudley Ingraham Papers, III, 33; Ingraham Company, "Age Groups in the E. Ingraham Employ March 7, 1940," Ingraham Company Papers; interview with Frank Savage by author, April 16, 1997. On "light manufacturing elsewhere," see Sharon Hartman Strom, *Beyond the Typewriter: Gender, Class, and the Origins of Modern Office Work, 1900-1930* (Urbana: University of Illinois Press, 1992).

1.40. Ingraham Company, untitled report, ca. 1921, Ingraham Company Papers, III, 12; description of Clock Manufacturers' Association of America, 1941, Edward Ingraham Papers, V, 26; W. K. Sessions to Mr. John Ivory, May 15, 1941, Edward Ingraham Papers, I, 3; *Bristol Press*, June 3, 1940.

1.41. Santago, Giammateo, and Giammateo interview.

1.42. Interview with Dorothy "Dottie" Beaucar by author, October 9, 1996.

1.43. Interview with Peter Burns by author, November 20, 1996.

1.44. Interviews by author with Rose Calderoni, December 4, 1996; George Power, October 15, 1996; Ingraham Company, "Employees who will have served during 1947 a total of 25 years or more but less than 50 years," June 2, 1947, Edward Ingraham Papers, V, 26; *Bristol Press*, October 25 and 30, 1951, Nov. 6, 1952, November 6, 1953, November 5, 1954, November 18, 1955.

1.45. Edward Ingraham, "A Personal Glimpse," 154; author's conversation with Kay Laviero, Dottie Beaucar, and Rosa Carrasquillo, May 29, 2000.

1.46. Conversation with Laviero, Beaucar, and Carrasquillo; Tetro interview.

1.47. Author's reading of *Bristol, Plainville, and Terryville Directory* (New Haven: Price & Lee), 1925, 1930, 1934, 1940, 1950, 1952, 1960; Beaucar, Webster interviews; interview with Nelson Spring by author and Dottie Beaucar, October 18, 1996. Susan Porter Benson outlines how families negotiated among their own members and with the market as well in *Household Accounts: Working-Class Family Economies in the Interwar United States* (Ithaca, NY: Cornell University Press, 2008). Particularly useful is the chapter entitled "Cooperative Conflict: Gender, Generation, and Consumption in Working-Class Families," which explores relations between parents and children in a consumer-oriented economy.

1.48. Cieszynski interview; interview by author with Walantyna Sakowski, November 22, 1996. Lake Compounce, among the largest parks in the area, frequently hosted Big Bands during the 1930s through the early 1950s, making it particularly popular among local workers.

1.49. Interview with Jerry Nocera by author, April 8, 1997; Laviero and Budnik interview.

1.50. Spring and Nocera interviews.

1.51. Laviero and Budnik interview. While occurring in other ethnic groups, Kay Laviero and Jerry Nocera's examples have added weight here within the context of the ethnicity that fostered them. Each instance represents a change in the traditional authority that fathers practiced over their children in household relations transplanted from southern Italy by immigrant parents. In the rural peasantry of that region, socioeconomically characterized by precapitalistic production where male household heads predominated, fathers exerted unquestionable control in the lives of family members. In the urban capitalist setting of Bristol, however, this authority waned, giving way to greater power for offspring in immigrant households. Judith E. Smith recognizes this breakdown happening among Italian American families elsewhere as well in the first half of the twentieth century, detailing it in Providence, RI. See Smith's *Family Connections: A History of Italian and Jewish Immigrant Lives in Providence, Rhode Island, 1900-1940* (Albany, NY: State University of New York Press, 1985).

1.52. Interviews by author with Umberto "Al" Calderoni, December 4, 1996; Rose Calderoni; Laviero and Budnik.

1.53. Al Calderoni, Rose Calderoni, and Cieszynski interviews; *Bristol Press,* November 25, 1912.

1.54. Interview with "Bertha D."

1.55. Interview with John Denehy by August Erling, April 2, 1988, Voices of Bristol Oral History Project, Bristol Public Library, Bristol, CT. On vibrating and the history of homework, see Edward Ingraham, "Industrial Homework, August 24, 1962," Paper 13, "Reminiscences of Edward Ingraham, Clockmaker," Edward Ingraham Papers, IV.

1.56. *Bristol Directory,* 1940, 1944, 1952, 1960; Power interview.

1.57. Santago, Giammateo, and Giammateo interview.

1.58. *Ibid.*; Laviero and Budnik interview.

1.59. Interview with Philomena Carrier by author, May 12, 1997.

1.60. Interviews by author with Lillian Rock, October 23, 1996; Agnes Moquin, November 8, 1996.

1.61. Interviews with Baldaccini and Mongillo; Rose Calderoni. Examples of community building among married women elsewhere can be had through Ardis Cameron, *Radicals of the Worst Sort: Laboring Women in Lawrence, Massachusetts, 1860-1912* (Champaign, IL: University of Illinois Press, 1994) and Jacquelyn Dowd Hall, James Leloudis, Robert Korstad, Mary Murphy, LuAnn Jones, and Christopher B. Daly, *Like a Family: The Making of a Southern Cotton Mill World* (Chapel Hill: University of North Carolina Press, 1987).

1.62. Baldaccini and Mongillo interview.

1.63. Richard Sennett, *Authority* (New York: Random House, 1980), 51.

1.64. *Bristol Press,* April 17, 1923; Dudley Ingraham to William S. Ingraham, circa April 16, 1923, Ingraham Company Papers, II, 11; Bristol labor turnover report for May 1923; State of Connecticut Public Document No. 23: *Thirty-First Report of the Bureau of Labor Statistics for the Period Ended June 30th, 1924* (Hartford, 1925); Clouette and Roth, Bristol, 183. On the IWW in Bristol, see *Bristol*, 183-184.

1.65. Edward Ingraham, "Personal Glimpse," 156; interviews with Laviero and Budnik; and Nocera.

1.66. Edward Ingraham, "The Ingraham Company in World War I and II," Paper VIII of "Reminiscences," May 9, 1962, 8; Edward Ingraham Papers, IV, 22; *Bristol Press,* December 7, 1951.

1.67 Joseph Ingraham to Dudley Ingraham, July 29, 1961, Dudley Ingraham Papers; *Bristol Press,* April 9, 1954 and August 23, 1956; Seymour M. Ingraham to Dudley Ingraham, December 12, 1953, author's collection; Edward Ingraham, "Personal Glimpse," 156; author's telephone conversation with David W. Sessions, November 18, 1996.

Chapter 2

2.1. Interview with Carl Kirkby by author, October 23, 1996.

2.2. "The E. Ingraham Company," October 19, 1938, Edward Ingraham Papers.

2.3. *Ibid.*; interviews with Beaucar; Laviero and Budnik; and Carrier.

2.4. George Power interview.

2.5. Tetro interview.

2.6. Ingraham Company memo, September 5, 1936, Ingraham Company Papers, VI; Clock Manufacturers' Association of America, "Quick Survey of Present Conditions," November 4, 1930, Ingraham Company Papers, III, 12.

2.7. Edward Ingraham, "Ingraham Watch Department Notes," *Timepiece Journal*, No. 5 (Winter 1996): 163-164, 163; interview with John Denehy by author, October 2, 1996; Ingraham Company/Dudley Ingraham to Mr. Edward Ingraham, November 23, 1933, Ingraham Company Papers, VI, 292.

2.8. Denehy interview by Erling.

2.9. *Ibid.*

2.10. Denehy interview by author.

2.11. Kirkby interview.

2.12. Interview by author with anonymous man, "Bill V," October 30, 1996; Kirkby interview.

2.13. "Bill V." interview.

2.14. *Bristol Press*, June 30, 1954; Kirkby interview.

2.15. Kirby interview.

2.16. Tetro interview.

2.17. *Bristol Press,* June 30, 1954; interview with August Erling by author, November 8, 1996.

2.18. Ingraham Company, "Comparison of Rates, Automatic Screw Machine," February 20, 1943. Ingraham Company Papers, VI, 290.

2.19. Ingraham Company, "Study of Labor Rates, Woodworking Industry, 1937," August 9, 1937, Ingraham Company Papers, III, 12.

2.20. Interview with Oreste DePascale by author, October 18, 1996.

2.21. Interview with Thomas "Tommy" DiSabato, April 2, 1997; Nocera interview.

2.22. Nocera interview.

2.23. Denehy interview by Erling.

2.24. Denehy interview by author.

2.25. Beaucar and Denehy interviews; "Ingraham Watch Department Notes"; Denehy interview by Erling. In contrast, women on the large men's wristwatches performed about two-thirds of the production of pocket watches, and the adjuster about three times the work he did was on pocket watches.

2.26. Carrier interview.

2.27. Santago, Giammateo, and Giammateo interview.

2.28. Laviero and Budnik interview.

2.29. Interviews with Rose Calderoni; Rock.

2.30. Ingraham office memo, circa February 19, 1927; Ingraham Company memo on compensation, both Dudley Ingraham Papers, III, 33.

2.31. Carrier interview; Patricia Cooper, *Once a Cigar Worker: Men, Women, and Work Culture in American Cigar Factories, 1900-1919* (Urbana: University of Illinois Press, 1987). Radium dials remained popular through the early 1960s when organizations like the New York City Board of Health banned sales of radium pocket watches, which the latter did as of February 1, 1963. Source: *Bristol Press,* November 27, 1962. On Ingraham's radium department, see Philip Samponaro, "Painting a Nightmare: Women and Radium Work at the E. Ingraham Company, 1923-1935," *Labor: Studies in Working-Class History of the Americas,* No. 4 (2) (Summer 2007): 9-22.

2.32. Interview by author with Domenick Dellario,

March 18, 1997; interviews with Moquin; Baldaccini and Mongillo.

2.33. Interviews with Santago, Giammateo, and Giammateo.

2.34. Cieszynski and Moquin interviews.

2.35. Laviero and Budnik interview.

2.36. *Ibid.*; Service Department/Wm. Large to Edward Ingraham, May 5, 1943, Ingraham Company Papers, VI, 291.

2.37. Interview with Denehy by Erling; Nocera interview.

2.38. Denehy and Nocera interviews.

2.39. Interview with Mary Penoncello by author, March 17, 1997; Laviero and Budnik interview; Baldaccini and Mongillo interview.

2.40. Author's interview with Patricia Letizia, December 4, 1996; undercover reports by "X127," or W. W. M., on working conditions and authority in the service department, December 1947-April 1948, Ingraham Company Papers, I, 7.

2.41. Interviews with anonymous woman and Kirkby.

2.42. Kirkby interview.

2.43. Carrier interview.

2.44. Letizia, Carrier, Sakowski, and Rose Calderoni interviews; interview by author with Frances Avery, June 5, 1997.

2.45. Sakowski interview.

2.46. *Ibid.*; KL interview. In practical terms, individual workers on fuses performed one of several operations, all of which had close counterparts in watchmaking. The first involved blanking plates out, including all holes. Shaving the plates to required tolerance followed, and the sanding and grinding that gave plates a finish thereafter. The fourth operation encompassed burring, reaming, threading, and countersinking, as well as drilling holes on the side of the plates for the screws to mount the plates and further piercing holes. In this latter series of procedures, certain plates also had to be staked together.

2.47. Ingraham Company, "Notice to Employees," May 1, 1942, Ingraham Company Papers, 291; and memo on employees not considered citizens, circa early 1942, Ingraham Company Papers, 292; UE Local 260, grievance on "alien angle," July 2, 1942, and Edward Ingraham to Carl H. Miller, July 10, 1942, both Edward Ingraham Papers, I, 4; Al Calderoni interview.

2.48. Kirkby interview.

2.49. *Ibid.*

Chapter 3

3.1. On the UE and its famed "them and us stand," see James Matles and James Higgins, *Them and Us: Struggles of a Rank and File Union* (Englewood Cliffs, NJ: Prentice-Hall, 1974). A useful counterpoint to understanding the broad story of a CIO union that Matles presents is Charles Williams' "Reconsidering CIO Political Culture: Briggs Local 212 and the Sources of Militancy in the Early UAW, *Labor: Studies in Working-Class History of the Americas,*" No. 7 (4) (Winter 2010): 17-43. Studying the example of a United Auto Workers local in Detroit, Williams acknowledges that an evolving New Deal political culture informed labor radicalism nationwide. At the same time, his analysis recognizes the relevance to the CIO of "a distinct form of working-class consciousness rooted in local experiences and workplace relationships" (19). Williams contends that this local consciousness must be underscored to comprehend the CIO's full story in line with my findings here.

3.2. *Bristol Press*, April 17, 26, May 2, 15, 1917; June 5, 1918; July 8, 1937; December 16, 17, 1938.

3.3. *Ibid.*, December 16, 17, 1938.

3.4. Ingraham Organizing Committee to Workers of Ingraham, June 6, 1940, Edward Ingraham Papers; UE Local 264 (Bristol), "Application for Charter," Organization 1941, FF731, UE Archive, the University of Pittsburgh.

3.5. Joseph Caiazza to UE, "Application for Position as International Representative or Field Organizer," June 28, 1940; Matles to Caiazza, July 1, 1940; Joseph Caiazza to James Matles, December 9, 1940 and June 7, 1941, Organizers Reports/Files, FF1a and FF1b; Charlie Rivers to Matles, June 27, 1940, Organizers Reports/Files, F1172, UE Archive.

3.6. Denehy interview; telephone conversation by author with David Rosenberg of the UE Archive, Storrs, CT, to Pittsburgh, fall 1996.

3.7. Laviero and Budnik interview.

3.8. Nocera interview.

3.9. Savage interview.

3.10. Nocera and Savage interviews.

3.11. Nocera interview; *Bristol Press,* April 11, 1941.

3.12. *Bristol Press,* March 27, 13, 1941.

3.13. "From Edward Ingraham, President, and D. S. Ingraham, Vice President," March 1941, Ingraham Company Papers, VI, 291.

3.14. Denehy and Tetro interviews: UE Local 260, "Officer Lists, 1941-1949," FF682, UE Archive; Edward Ingraham, "Reminiscences of Edward Ingraham, Clockmaker," Edward Ingraham Papers. Officer lists for UE Local 260 show DeCapua's first election as union president on August 17, 1942. DeCapua thereafter appears as president—and shop steward for movement shop as well—consecutively from December 1943 through 1949 and the IUE-UE crisis at the plant.

3.15. George Frederickson, "Analysis of the Trends of the E. Ingraham Company with Suggestions as to Future Policies in Order to Insure Continued Growth and Success," September 24, 1937, 12, Edward Ingraham Papers I, 1; Payroll Ledger of the E. Ingraham Company, First Quarter, 1942, Ingraham Company Papers; DiSabato interview. The cabinet and sanding department employed about one-third of all case shop workers.

3.16. DePascale interview.

3.17. Employees of the Case Shop, E. Ingraham Company, to Edward Ingraham, March 17, 1940; memo by

Edward Ingraham, May 5, 1940, both Edward Ingraham Papers.

3.18. DiSabato interview.

3.19. Author's reading of the *Bristol, Plainville and Terryville Directory, 1940* (New Haven: Price & Lee Company, 1940) for sample of Italian Americans in the case shop and their family status; DePascale and DiSabato interviews.

3.20. *Bristol Press*, July 7, December 17, 1955; Tetro interview. Ingraham regularly turned to Victor in securing bosses for its case shop; Ciccarelli's immediate successor, Charles Stitt, worked at Victor before taking the job of case shop superintendent at Ingraham as of August 1926.

3.21. Ingraham, "Reminiscences," IV, 12; DiSabato interview; Dudley Ingraham to Elof Carlson, May 6, 1940, Edward Ingraham Papers; comments by Kay Laviero to author, January 19, 2003. DiLorenzo's strong activism in the case shop won him election as UE 260's first president both before the NLRB victory in March 1941 and then again on July 24, 1941. As president, DiLorenzo also sat on the union's first negotiating committee.

3.22. DePascale and DiSabato interviews; Edward Ingraham, "A Personal Glimpse," 156.

3.23. DePascale and DiSabato interviews; report by Dudley Ingraham. In fact, Capone did take the job, at least for a period; company records list him as foreman of the finishing department in the late 1930s.

3.24. DiSabato interview.

3.25. Edward Ingraham, "Personal Glimpse," 156; Kirkby and Tetro interviews; Edward Ingraham, memorandum to Mr. Elof Carlson, February 21, 1941, Edward Ingraham Papers.

3.26. Rock interview.

3.27. "Ingraham Union News," November-February/March, 1941, Ingraham Company Papers, VI, 290.

3.28. "Independent Union," March 24 and 25, 1941, Ingraham Company Papers, VI, 290.

3.29. Denehy interview.

3.30. Tetro and Kirkby interview.

3.31. A satisfied employee to Mr. Ingraham, March 26, 1941, Edward Ingraham Papers.

3.32. *Ibid.*

3.33. *Ibid.*

3.34. Laviero interview.

3.35. Ingraham Clock Workers Building Union," *UE News,* December 14, 1940; "Ingraham Union News," circa October 15, November 7, circa November 10-24, 1940, January 1941, Ingraham Company Papers, VI, 290; Wendell Copeland to UE, July 25, 1941, Local 260, "Officer Lists, 1941-1949," FF682; UE Local 264, "Application for a Local Union Charter," FF731, "Organization, 1941"; and Josephine Forcella to UER&MWA, September 30, 1941; Helen Fisher to Julius Emspak, June 18 and July 23, 1942, both "Plant Closure, 1942-1943," FF732, UE Archive.

3.36. Elizabeth Faue, *Community of Suffering and Struggle: Women, Men, and the Labor Movement in Minneapolis, 1915-1945* (Chapel Hill: University of North Carolina Press, 1991), 183. Looking at a century-long shift from community-based to workplace activism, Faue determines that women in the CIO campaigns of 1930s Minneapolis encountered the resilience of familial and political ideology that separated them from the workforce. She shows that, while women were key participants in the community-based union activism of the 1930s, a bureaucratic unionism in place by the end of the decade viewed them with increased hostility and generally excluded them from the "symbolic system of labor culture" because the duration of the Depression had focused attention on men due to their perceived status as breadwinners. Faue points out that the regendering of the workforce with the coming war effort caused unions to be more responsive to women, again reminding us that the gendering of solidarity is malleable, depending on circumstance, industry, and community.

3.37. Tetro interview.

3.38. *Bristol Press*, May 16, 17; August 12, 1941, "Ingraham Union News," circa August 1-September 26, 1941, Ingraham Company Papers, VI, 290.

3.39. Power interview.

3.40. "From Edward Ingraham, President, and Dudley S. Ingraham, Vice President," Tetro interview.

3.41. Laviero and Budnik interview.

3.42. Denehy and Carrier interviews.

3.43. Spring interview.

3.44. DePascale interview.

3.45. George DuBois to Julius Emspak, March 18, 1942; Emspak to DuBois, March 23, 1942, both UE 264, "Correspondence-Incoming," FF728, UE Archive; Helen Fisher to Julius Emspak, June 18 and July 23, 1942, FF732.

3.46. *Bristol Press,* June 16, 1944.

3.47. Walter W. Cenerazzo, National President, to All Ingraham Clock Employees, March 18, 1946, Ingraham Company Papers, 292; Bruce Waybur to Albert DeCapua, May 15, 17, 1946, and Dudley Ingraham to Albert DeCapua, May 21, 23, 1946, Ingraham Company Papers, III, 128.

3.48. *Albert DeCapua, Etc.* vs. *E. Ingraham Company,* January 4, 1947, U. S. District Court, District of Connecticut, Civil Action #1976, Ingraham Company Papers, 292; *Hartford Courant*, January 1, 15, 1947; *Bristol Press*, January 9, 10, 14, 1947; Ingraham Company to Our Employees, January 8, 1947, Ingraham Company Papers, I; Edward Ingraham, memorandum, January 14, 1947, Edward Ingraham Papers, I, 6.

3.49. *Bristol Press,* January 10, 1947; Reorganization Committee, "DEMOCRACY 'One for All? Or All for One!!!" circa January 30, 1947, Ingraham Company Papers, VI, 290.

3.50. Edward Ingraham, "Truth Can Keep US Free," *Connecticut Industry* (September 1947): 7, and "Danger Signals," *Connecticut Industry* (October 1947): 7; *Bristol Press*, October 14, 1948; Ronald L. Filipelli and Mark D. McColloch, *Cold War in the Working Class: The Rise and Decline of the United Electrical Workers* (Albany: State University of New York Press, 1995), 141; Officers and Stewards of UE Local 260, "Answer These Questions Mr. Flanagan," November 18, 1949, Ingraham Company Papers, VI, 290. For localized accounts of the UE-IUE crisis, see Robert W. Cherny, William Issel, Kiernan Walsh Taylor, eds., *American Labor and the Cold War: Grassroots Politics and Postwar Political Culture* (New Brunswick: Rutgers University Press, 2004).

3.51. Executive Boards, UE Locals 260 and 256, "Statement on Hartford Secession," "Leaflets 1947," FF 689, UE Archive; UE Local 260, "UAW Raids Don't Pay Off," *Clockwise* (February 1949): 2, 4, Ingraham Company Papers, VI, 290; Filipelli and McColloch, *Cold War*, 118-119; Organizing Committee, IUE-CIO, "Your Future Is at Stake Now!!!," November 14, 1949, Ingraham Company Papers, VI, 290.

3.52. Power interview.

3.53. Kirkby interview.

3.54. *Ibid*.

3.55. Moquin and Kirkby interviews.

3.56. *Bristol Press,* November 14, 1949.

3.57. Local 260 IUE-CIO, "Is UE Communistic?," February 1, 1950; "Bulletin," March 6, 1950; and "Election Ordered for Ingraham's," April 19, 1950, all Ingraham Company Papers, VI, 290.

3.58. UE Local 260, "Ingraham Workers Support UE Local 260," *Clockwise* (May 10, 1950): 3, Ingraham Company Papers, VI, 290; *Bristol Press*, May 12, 1950; IUE 260, "Election Results," May 12, 1950; UE Local 260, "UE Largest Union," May 31, 1950, both Ingraham Company Papers, VI, 290; "UE Local 260 Radio Broadcast, May 31, 1950," 5, Beach, Calder, Anderson, and Alden Papers, Box 20, Archives and Special Collections, University of Connecticut, Storrs.

3.59. UE Local 260, "THERE IS NO END TO THE LIES OF THE IUE-CIO UNION WRECKERS," June 8, 1950, Ingraham Company Papers, VI, 290; Kirkby interview; *Bristol Press*, June 8, 9, 1950.

3.60. Ingraham Company, "Notice," June 13, 1950, and IUE Local 260, "Introducing the 'Tick Tock,'" June 1950, both Ingraham Company Papers, VI, 290; "Agreement between the E. Ingraham Company and the International Union of Electrical, Radio and Machine Workers, CIO, Local 260, September 14, 1950," author's collection; Edward Ingraham, "Reminiscences," IV, 13; *Bristol Press*, September 14, 1950, April 10, 1951; Martha R. Helming, Clock Number 1935, to Ingraham Company, June 14, 1950, Ingraham Company Papers, VI, 291. For unclear reasons, Flanagan had only a brief stint as IUE Local 260's president; IUE elections during the summer gave the office to Almeda Page, who as Flanagan's successor, signed the five-year contract at its conclusion in September.

3.61. Power and Letizia interviews; *Bristol Press*, October 15, 1958.

Chapter 4

4.1. *Bristol Press*, July 1, 8, 16, 18, 20, and 25, 1960.

4.2. Clouette and Roth, *Bristol, Connecticut*, 241.

4.3. *Bristol Press*, January 19, March 20, September 24, 1957; April 1, 1958; December 12, December 29, 1960; April 17, 1964; March 19, September 20, 1965; May 4, 1966; May 5, 1967; Consolidated Electronics Industries Corporation by R. G. Dettmer to Mr. William K. Sessions, Jr., January 21, 1957; Sessions Clock Company by W. K. Sessions to All Creditors of the Sessions Clock Company, February 18, 1957; Sessions Clock Company, Balance Sheet, December 31, 1956, all Dudley Ingraham Papers, I, 30; A. W. Haydon, "President's Report to the Stockholders" for the years ended December 31, 1959, December 31, 1961, American Clock and Watch Museum; Manufacturers Association of Hartford County, Inc., summary labor turnover report for Bristol District, November 16, 1955; Ingraham Company Papers, VI, 290.

4.4. Chris Bailey, "The End of an Era: The Sessions Clock Company," 114; Kneeland, "Sessions Clock," 8-9; Dellario interview; *Bristol Press*, September 20, 1965.

4.5. Dellario interview; *Bristol Press*, July 1, 1965.

4.6. Tetro, Spring, and Al Calderoni interviews.

4.7. *Bristol Press*, September 5, and October 23, 1957.

4.8. Edward Ingraham, "A Personal Glimpse," 154; *Bristol Press*, December 3, 1962, April 25, 1963.

4.9. Ingraham, "Personal Glimpse," 161. Ingraham had already shifted to its gaze to Appalachia and its vicinity earlier in the decade when opening up a timer motor assembly plant in Elizabethtown, KY, in 1954 at the request of P.M. Mallory, a major client who wanted a second site to ensure adequate production. As implicit in Mallory's desire, the Elizabethtown plant—often referred to as "E-town"—meant to supplement, and not undermine, the Bristol workforce.

4.10. Spring interview; Ingraham, "Personal Glimpse," 161.

4.11. *Bristol Press*, October 29, December 17, 19, 22, 1958; January 23, 31, February 3, 5, 10, 1959; July 1, 20, 25, 1960; IUE Local 260 and Ingraham Company, union agreement of August 31, 1960; Ingraham Company Papers, VI, 290.

4.12. Sakowski interview; *Bristol Press*, March 15, 1991; Dorothy Beaucar, "Bussmann Employees—Years of Service," March 18, 1986; revised list of Bussmann employees by grade classification, September 29, 1976; and "Bussmann – Commercial Development," 1977, all courtesy of Dorothy Beaucar.

4.13. Between February 1957 and the company's last shipment in December 1973, the firm fulfilled 15 orders for the "M125" booster, and beginning with a 2.9 million-dollar order at the end of June 1961 and ending in January 1974, received 15 orders for traditional fuses as well. Source: *Bristol Press*, September 18, 1957; April 1, 1960; July 16, 1960; June 29, 1961; Dorothy Beaucar, "Fuze Contracts, Shipments Per Month," and "M125A1

Booster Contracts," both courtesy of Beaucar.

4.14. Rock interview.

4.15. Interview with Georgia Parlyak by author, September 25, 1996.

4.16. *Ibid.*

4.17. *Ibid.*

4.18. Victor Perlo, "Social Effects of the Science-Technological Revolution," *Political Affairs* (August 1964): 20-35, 22-23, 29.

4.19. *Ibid.*, 29.

4.20. Interview with "Bill V." by author, October 30, 1996.

4.21. Stanley Ruttenberg in U.S. Bureau of Labor Statistics, *Monthly Labor Review* (February 1959).

4.22. Erling interview.

4.23. *Ibid.* As ordnance programs "were slowing down" in the early 1970s, the firm reassigned Augie to other areas of work. "Then I moved back into some of the other type work that Ingraham was doing with the electric motors and timers and that sort of thing." Thereafter, Augie moved into quality control and served as product engineering manager until a major restructuring after the evaporation of all contracts in 1974 cost him his job.

4.24. "Bill V." interview.

4.25. *Ibid.*

4.26. Sakowski interview. Alice Kessler-Harris chronicles women's changing relationship with wage-earning in her *In Pursuit of Equity: Women, Men, and the Quest for Economic Citizenship in 20th Century America* (New York: Oxford University Press, 2001).

4.27. Steve Meyer offers a similar thesis on the confusion of class boundaries by arguing that outside consumption as a result of wage-earning gave a sense of respectability to workers who otherwise participated in a rough shop-floor culture. Consult Meyer's essay on this subject in Roger Horowitz, editor, *Boys and Their Toys?: Masculinity, Technology, and Class in America* (New York: Routledge, 2001).

4.28. *Bristol Press,* June 15, 1957; October 31, 1966.

4.29. *Ibid.*, December 21, 1966.

4.30. *Ibid.*, May 27, 1967; Dennis Deslippe, *"Rights Not Roses": Unions and the Rise of Working-Class Feminism, 1945-1980* (Champaign: University of Illinois Press, 2000).

4.31. George Power interview.

4.32. Period reports on Sessions employees and "The Sessions Clock Company, Actuarial Valuation of the Pension Plan As of August 1, 1969," all Beach, Calder, Anderson, and Alden Papers, 16; 1960 "sample" based on author's reading of *Bristol, Plainville, Terryville Directory, 1960* (New Haven: Price & Lee, 1960); *Bristol Press,* May 16, 1967. The 1960 sample revealed 56 percent of the women with husbands and another 10 percent as widows.

4.33. *Bristol Press,* May 5-July 11, 1967; interviews by author with Janet Watson, March 17, 1997; Frances Avery, June 5, 1997; Cieszynski, Santago, Giammateo, and Giammateo; Carrier; Baldaccini and Mongillo.

4.34. Baldaccini and Mongillo interview.

4.35. *Bristol Press,* May 12, 16, 20, 1967; Baldaccini and Mongillo interview.

4.36. *Bristol Press,* May 20, October 11, 1967.

4.37. *Ibid.,* May 20, 1967; Baldaccini and Mongillo interview. State action against James Ingalls, his wife Pauline, and Clayton Aiudi for breach of the peace began in late September 1967. Several trials ensued over the next four years and ended with the dismissal of all charges on September 24, 1971. Meanwhile, the three defendants, with injured striker Leona Sears, sued several members of the Police Department and the City of Bristol in November 1967. On the legal proceedings associated with the strike, see *Bristol Press,* September 30, October 10, 11, 12, November 16, 1967; February 26, 27, 1969; September 24, 1971; *Hartford Courant,* February 25, 26, 28, 1969; May 19, 1973; folder titled "Court Case Feb. 1969 IUE-AFL-CIO Local 261," James Ingalls Papers, Box 6A, Archives and Special Collections, University of Connecticut, Storrs, CT.

4.38. Interviews with Cieszynski; Watson; Dellario; Santago, Giammateo, and Giammateo.

4.39. *Bristol Press,* May 24, 25, July 8 and 11, 1967; draft of agreement between the Sessions Clock Company and IUE Local 261, AFL-CIO, August 29, 1967, Beach, Calder, Anderson, and Alden Papers, 16.

4.40. Interviews with Dellario, Cieszynski, Shores; *Bristol Press,* October 1, 1968, April and August 15, 1969; March 12, May 1 and 5, 1970; Bailey, "The End of an Era: The Sessions Clock Company," 14.

4.41. Baldaccini and Mongillo interview.

4.42. *Ibid.*

4.43. Interviews with Cieszynski; Santago, Giammateo, and Giammateo; Shores.

4.44. Interviews with Baldaccini and Mongillo; Cieszynski.

Epilogue

E.1. Baldaccini and Mongillo interview.

E.2. Power interview

E.3. "The Defense Department Makes a Timely Purchase," *Business Week* (August 14, 1971): 27; Dudley S. Ingraham to Ella T. Grasso, July 16, 1971; and Ella Grasso to Dudley Ingraham, August 11, 1971, author's collection.

E.4. Power interview.

E.5. *Ibid.*

Bibliography

Manuscript Collections

Beach, Calder, Anderson, and Alden Papers, University of Connecticut.

Dudley S. Ingraham Papers, University of Connecticut.

E. Ingraham Company Papers, University of Connecticut.

Edward Ingraham Library, American Clock and Watch Museum, Bristol, CT.

Edward Ingraham Papers, University of Connecticut.

Ingraham family papers, author's collection.

James Ingalls Papers, University of Connecticut.

Microfilm of manuscript returns of CT censuses, 1880-1920, University of Connecticut.

Reconstruction Finance Corporation, Record Group 234, National Archives.

UE Archive, University of Pittsburgh.

Voices of Bristol Oral History Project, Bristol Public Library.

Newspapers and Periodicals

Bristol, Plainville and Terryville Directory

Bristol Herald

Bristol Press

Connecticut Industry

Hartford Courant

Monthly Labor Review

Political Affairs

Timepiece Journal

UE News

Interviews by Author

Avery, Frances. June 5, 1997. Telephone.

Baldaccini, Catherine "Kitty," and Hazel Mongillo. June 4, 1997. Southington, CT.

Beaucar, Dorothy "Dottie." October 9, 1996. Bristol, CT.

"Bertha D." (Anonymous). March 17, 1997. Bristol, CT.

"Bill R." (Anonymous). October 30, 1996. Bristol, CT.

Burns, Peter. November 20, 1996. Bristol, CT.

Calderoni, Rose. December 4, 1996. Bristol CT. With Umberto Calderoni.

Calderoni, Umberto. December 4, 1996. Bristol, CT.

Carrier, Philomena "Phil." May 7, 1997. Bristol, CT.

Cieszynski, Lois. March 18, 1996. Forestville, CT.

Denehy, John. October 2, 1996. Bristol, CT.

Dellario, Domenick. March 18, 1997. Wolcott, CT.

DePascale, Oreste. October 18, 1996. Bristol, CT.

DiSabato, Thomas "Tommy." April 2, 1997. Bristol, CT.

Erling, August "Augie." November 8, 1996. Bristol, CT.

Kirkby, Carl. October 23, 1996. Bristol, CT.

Kondziolka, Walter. October 23, 1996. Bristol, CT.

Laviero, Kathryn "Kay," and Mary Budnik. October 16, 1996. Bristol, CT.

Letizia, Patricia "Pat." December 4, 1996. Bristol, CT.

Moquin, Agnes. November 8, 1996. Bristol, CT.

Nocera, Jerry. April 8, 1997. Forestville, CT.

Parlyak, Georgia May. September 25, 1996. Unionville, CT.

Penoncello, Mary. March 17, 1997. Bristol, CT.

Power, George. October 15, 1996. Bristol, CT.

Richardson, Barbara. December 4, 1996. Bristol, CT.

Riquier, Margaret Munn. March 26, 1997. Bristol, CT.

Rock, Lillian. October 23, 1996. Bristol, CT.

Russell, Carolyn. March 26, 1997. Forestville, CT.

Sakowski, Walantyna. November 22, 1996. Bristol, CT.

Santago, Laura, Sandy Giammateo, and Domenic Giammateo. April 2, 1997. Southington, CT.

Savage, Francis "Frank." April 16, 1997. Plainville, CT.

Sessions, David W. November 18, 1996. Telephone.

Shores, Frederick. April 25, 1997. East Hartford, CT.

Spring, Nelson. October 18, 1996. Bristol, CT. With Dorothy Beaucar.

Tetro, Robert. October 30, 1996. Bristol, CT.

Watson, Janet. March 17, 1997. Forestville, CT.

Webster, Kenneth. October 30, 1996. Bristol, CT.

Follow-up Interviews

Beaucar, Dorothy "Dottie," and Kay Laviero. May 29, 2000. Bristol, CT. With Rosa E. Carrasquillo.

Beaucar, Dorothy "Dottie," and Kay Laviero. February 9, 2002. Bristol, CT.

Books and Articles

Bailey, Chris H. "Sessions' Employee Numbers, 1930-1953," and "The End of an Era: The Sessions Clock Company." Sessions Clocks Catalogue No. 65, 1915. Bristol: American Clock and Watch Museum, 1977.

Bailey, Chris H. Fifty Years of Time: The First 50 Years of the American Clock and Watch Museum. Bristol: American Clock and Watch Museum, 2009.

Barr, Lockwood. "Clockmakers of Bristol." Unpublished manuscript in E. Ingraham Company Papers at the University of Connecticut.

Beals, Carleton. The Making of Bristol: Our Yankee Heritage. Bristol: Bristol Public Library, 1954.

Benson, Susan Porter. Counter Cultures: Saleswomen, Managers, and Customers in American Department Stores, 1890-1940. Urbana: University of Illinois Press, 1987.

Benson, Susan Porter. Household Accounts: Working-Class Family Economies in the Interwar United States. Ithaca, NY: Cornell University Press, 2008.

Brecher, Jeremy, Jerry Lombardi, and Jan Stackhouse-which. Brass Valley: The Story of Working People's Lives and Struggles in an American Industrial Region. Philadelphia: Temple University Press, 1982.

Breugel, Martin. "'Time That Can Be Relied Upon': The Evolution of Time Consciousness in the Mid-Hudson Valley, 1780-1860." Journal of Social History (Spring 1995): 547-564.

(Bristol) World War II Historical Committee. Bristol, Connecticut in World War II . . . Depicting the Part Played by Our Brave Men and Women Who Dedicated Their Services to Their Country, As Well As the Mobilization of Our Home Life and Industries in Support of the War Effort, 1939-1946. Worcester, MA, 1947.

Cameron, Ardis. Radicals of the Worst Sort: Laboring Women in Lawrence, Massachusetts, 1860-1912. Urbana: University of Illinois Press, 1993.

Cherny, Robert W., William Issel, Kiernan Walsh Taylor, eds. American Labor and the Cold War: Grassroots Politics and Postwar Political Culture. New Brunswick: Rutgers University Press, 2004.

Clouette, Bruce, and Mathew Roth. Bristol, Connecticut: A Bicentennial History, 1785-1985. Canaan, NH: Phoenix Publishing, 1984.

Connecticut Bureau of Labor. Thirty-First Report of the Bureau of Labor Statistics for the Period Ended June 30th, 1924. Hartford, 1925.

Cooper, Patricia. Once a Cigar Maker: Men, Women, and Work Culture in American Cigar Factories, 1900-1919. Urbana: University of Illinois Press, 1987.

Deslippe, Dennis. "Rights Not Roses": Unions and the Rise of Working-Class Feminism, 1945-1980. Urbana: University of Illinois Press, 2000.

Dunaway, David King, and Willa K. Baum, eds. Oral History: An Interdisciplinary Anthology. Walnut Creek, CA: AltaMira Press, 1996.

Faue, Elizabeth. Community of Suffering and Struggle: Women, Men, and the Labor Movement in Minneapolis, 1915-1945. Chapel Hill: University of North Carolina Press, 1991.

Filipelli, Ronald L., and Mark D. McColloch. Cold War in the Working Class: The Rise and Fall of the United Electrical Workers. Albany: State University of New York Press, 1995.

Fine, Lisa M. The Story of Reo Joe: Work, Kin, and Community in Autotown, U.S.A. Philadelphia: Temple University Press, 2004.

Gluck, Sherna Berger, and Daphne Patai, eds. Women's Words: The Feminist Practice of Oral History. New York: Routledge, 1991.

Gramsci, Antonio. "Americanism and Fordism." Selections from the Prison Notebooks. Edited by Quinton Hoare and Geoffery Nowell Smith. New York: International Publishers Company, 1971.

Hall, James Leloudis, Robert Korstad, Mary Murphy, LuAnn Jones, and Christopher B. Daly. Like a Family: The Making of a Southern Cotton Mill World. Chapel Hill: University of North Carolina Press, 1987.

Hoke, Donald. Ingenious Yankees: The Rise of the American System of Manufactures in the Private Sector. New York: Columbia University Press, 1990.

Horowitz, Roger, ed. Boys and Their Toys?: Masculinity, Technology, and Class in America. New York: Routledge, 2001.

Ingraham, Edward. "A Personal Glimpse into the E. Ingraham Company and Clock Manufacturing in the 20th Century." Timepiece Journal, No. 5 (8) (Winter 1996): 151-162.

Ingraham, Edward. "Clockmaking in Connecticut." The Antiquarian: The Bulletin of the Antiquarian and Landmarks Society, Inc. of Connecticut, No. III (1) (June 1951): 10-16.

Ingraham, Edward. "Ingraham Watch Department Notes." Timepiece Journal, No. 5 (8) (Winter 1996): 163-164.

Ingraham Story, The. Bristol: The E. Ingraham Company, circa 1954.

Kessler-Harris, Alice. In Pursuit of Equity: Women, Men, and the Quest for Economic Citizenship in 20th Century America. New York: Oxford University Press, 2001.

Kneeland, Howard E. "The History of the Sessions Clock Company, 1829-1966." Typescript manuscript, courtesy of Francis Savage.

Kovaleff, Theodore. *Business and Government during the Eisenhower Administration: A Study of the Anti-Trust Division of the Justice Department.* Athens: Ohio State University Press, 1980.

Landes, David. *Revolution in Time: Clocks and the Making of the Modern World.* Cambridge, MA: Harvard University Press, Belknap University Press, 1983.

Ly, Tran Duy. *Ingraham Clocks and Watches.* Fairfax, VA: Arlington Books, 1998.

Ly, Tran Duy. *Sessions Clocks.* Fairfax, VA: Arlington Books, 2001.

Matles, James, and James Higgins. *Them and Us: Struggles of a Rank and File Union.* Englewood Cliffs, NJ: Prentice-Hall, 1974.

McCrossen, Alexis. *Marking Modern Times: A History of Clocks, Watches, and Other Timekeepers in American Life.* Chicago: University of Chicago Press, 2013.

McCrossen, Alexis. "The 'Very Delicate Construction' of Pocket Watches and Time Consciousness in the Nineteenth Century United States." *Winterthur Portfolio,* No. 44 (Spring 2010): 1.

Montgomery, David, et al. "Patronage, Paternalism, and Company Welfare." *International Labor and Working Class History Journal,* No. 53 (Spring 1998).

Murphy, John Joseph. "Establishment of the American Clock Industry: A Study in Entrepreneurial History." Unpublished Ph.D. dissertation, Yale University, 1961.

Mussey, Barrows. *Young Father Time: A Yankee Portrait.* New York: Newcomen Society in North America, 1950.

Nicholson, Linda. *Feminism/Postmodernism.* New York: Routledge, 1990.

Parr, Joy. *The Gender of Breadwinners: Women, Men, and Change in Two Industrial Towns, 1880-1950.* Toronto: University of Toronto Press, 1990.

Peck, Epaphroditus. *A History of Bristol, Connecticut.* Hartford: The Lewis Street Bookshop, 1932.

Perlo, Victor. "Social Effects of the Science-Technological Revolution," *Political Affairs* (August 1964): 20-35, 29.

Portelli, Allesandro. *The Death of Luigi Trastulli and Other Stories: Form and Meaning in Oral History.* Albany: State University of New York Press, 1990.

Rosenzweig, Roy. *Eight Hours for What We Will: Workers and Leisure in an Industrial City, 1870-1920.* Cambridge, UK: Cambridge University Press, 1983.

Samponaro, Philip. "Painting a Nightmare: Women and Radium Work at the E. Ingraham Company, 1923-1935." *Labor: Studies in Working-Class History of the Americas,* Vol. 4 (2) (Summer 2007): 9-22.

Samponaro, Philip. "Taxes, Tariffs, and Timepieces: International Trade and the Clock and Watch Industry in Bristol, CT, 1936-1955." *Connecticut History,* Vol. 48 (1) (Spring 2009): 1-29.

Samponaro, Philip. "Work, Gender, and the Twentieth-Century Workplace in Bristol's Clock and Watch Industry." *NAWCC Bulletin,* No. 376 (October 2008): 553-568.

Sessions Clock Company. 127 Years of Clock Craftsmanship. Bristol: Sessions Clock Company, ca. 1950.

State of Connecticut. *Public Document No. 23: Thirty-First Report of the Bureau of Labor Statistics for the Period Ended June 30th, 1924.* Hartford, 1925.

Sennett, Richard. *Authority.* New York: Random House, 1980.

Smith, Judith E. *Family Connections: A History of Italian and Jewish Immigrant Lives in Providence, Rhode Island, 1900-1940.* Albany, NY: State University of New York Press, 1985.

Strom, Sharon Hartman. *Beyond the Typewriter: Gender, Class, and the Origins of Modern Office Work, 1900-1930.* Urbana: University of Illinois Press, 1992.

Taylor, Snowden, and Kenneth D. Roberts. *Forestville Clockmakers.* Fitzwilliam, NH: Ken Roberts Publishing Company, 1992.

Turrini, Joseph M. "The Newton Steel Strike: A Watershed in the CIO's Failure to Organize 'Little Steel.'" *Labor History,* Vol. 38 (2-3) (Spring/Summer 1997): 229-265.

U.S. Department of Commerce. *Composition and Characteristics of the Population, 1910.* Washington, DC, 1913.

U.S. Department of Commerce. *Fifteenth Census of the United States: 1930, Unemployment, Volume II, General Report.* Washington, DC, 1933.

U.S. Department of Commerce. *Sixteenth Census of the United States: 1940; Manufactures: 1939, Vol. II, Part 2.* Washington, DC, 1942.

Williams, Charles. "Reconsidering CIO Political Culture: Briggs Local 212 and the Sources of Militancy in the Early UAW." *Labor: Studies in Working-Class History of the Americas,* Vol. 7 (4) (Winter 2010): 17-43.

Wingerd, Mary Lethert. "Rethinking Paternalism: Power and Parochialism in a Southern Mill Village." *Journal of American History,* No. 83 (December 1996): 872-902.

Zahavi, Gerald. *Workers, Managers, and Welfare Capitalism: The Shoeworkers and Tanners of Endicott Johnson, 1890-1950.* Urbana: University of Illinois Press, 1988.

Index

American Clock and Watch Museum, 3, 11, 15
American Tariff League, 13
American Watch Workers' Union, 52

Bailey, Chris A., 3, 57
Baldaccini, Catherine "Kitty," 16, 23, 35, 37, 66, 68, 71
Bannatyne Watch Company, 9
Bristol, 9, 10, 11, 13, 19, 21, 22, 23, 26, 30, 31, 36, 44, 45, 46, 48, 54, 63, 64, 71
    brassmaking in, 3, 7, 8, 42
    Chippins Hill neighborhood, 14
    civic, 7, 14, 15
    clockmaking locally, pre-1900, 7, 8, 14
    deindustrialization, 7, 69
    economy, 1, 2, 12, 13, 16, 20, 62, 72
    ethnic group(s), 2, 4, 17, 48
    industry, location of, 1, 2, 8, 57, 60, 72
    police, 55, 57, 66, 67, 68
    politics, 14, 24, 51, 53
    population, 8, 14, 17, 58
    unions, 24, 43, 50
Bristol Brass Company, 8
Bristol Press, 11, 15, 18, 44, 45, 52, 57, 66
Bulova, 71
Bussmann Division of McGraw-Edison, 59, 60

Caiazza, Joseph Louis, 43, 45
Ciccarelli, Ostilio, 46, 47, 48
Cieszynski, Lois, 20, 21, 35, 38, 67, 68
Civil Rights Act of 1964, 65
Clock Manufacturers Association of America, 15
Cold War, 4, 53, 55
    effects on Ingraham, 1, 13, 39, 43, 52, 62, 63, 71
    effects on Sessions, 43
Congress of Industrial Organizations (CIO), 1, 4, 7, 15, 17, 24, 25, 26, 43, 44, 46, 47, 49, 52, 53, 54, 55, 63
Consolidated Electronic Industries Corporation, 57, 66, 67
Cooper, Robert, 26, 57
    management style at Ingraham, 58, 59, 60

De Capua, Albert, 48, 53
Denehy, John, 22, 28, 29, 33, 37, 43, 45, 48, 51
Department of Defense, 71
DePascale, Oreste, 32, 45, 47, 51

Economic Opportunity Act of 1964, 65
Elgin Watch Company, 52
Emerson Radio and Phonograph Corporation, 15, 27, 46

Equal Pay Act of 1963, 65

Federal Excise Tax of 1941, 12
Flanagan, Richard C., 52, 53, 54, 55, 56
Fordism, 16
Forestville, 1, 7, 8, 14, 17, 20, 22, 25, 33, 54, 59, 66, 67, 71
Fuse-making (at Ingraham), 1, 11, 12, 13, 19, 40, 57, 59, 71
    engineers and, 25, 39, 62, 63, 64
    women's responses to, 23, 34, 60, 61, 62

Gilbert Clock Company, 16, 63
Gordon, Maurice, 58
Gramsci, Antonio, 16
Grasso, Ella, 71

Hamilton Watch Company, 52, 71
Haydon, Arthur, 11, 57, 58
Haydon Manufacturing Company, 11
"helping," worker practice of, 24, 38, 39, 40, 64
Hull Reciprocal Trade Agreements, 12

Industrial Workers of the World, 24
Ingalls, James, 67
Ingraham-Canadian Watch Company, Ltd., 45
Ingraham Company
    closing, 60, 65
    downsizing, 59, 60
    engineers, 62, 63
    fuses, 60, 61, 62, 63, 64
    Laurinburg (NC) assembly plant, 59, 60
    management, 58, 59
    McGraw-Edison sale, 59, 60
    Redstone Hill Road plant, 59, 60

Ingraham Company, E..
    automatic screw department, 12, 27, 31, 51
    case shop, 10, 15, 27, 31, 32, 33, 47, 48, 50, 51
    clocks, electric, 9, 11, 58
    clocks, manual, 9
    credit union, 22
    departments, see by name of department
    and Depression, 7, 10, 11, 15, 32
    division of labor, 1, 2, 17, 23, 27, 28, 31, 45
    early history, 7, 8,
    Elizabethtown (KY) assembly plant, 80, fn 4.9
    finishing department, 47
    foremen, 19, 28, 29, 47, 51
    fuses, antiaircraft, 11
    layoffs, 15, 19, 23, 46, 57, 64
    management, 2, 4, 57

paternalism, 2, 4, 7, 9, 13, 15, 17, 18, 19, 24, 25, 28, 29, 35, 38, 45, 47, 48, 49, 50
presswork, 34, 40
radio cabinets, 10, 15, 27, 45, 46
radium department, 27, 35
Sentinel Line, 25
service department, 37, 39
skilled labor, 18, 28, 30, 31, 32, 38, 49
strikes, 50, 52
Swiss watches' effects on, 12, 13
timers, 10
tool and die department, 18, 27, 29, 30, 31, 48
trucker(s), 27, 28, 38, 47
unions, 2, 32
unskilled labor, 18, 31, 32, 33
wages, 13, 15, 16, 24, 33, 40, 43, 46, 51, 63, 71
watch department, 19, 23, 27, 28, 29, 33, 34, 38, 51
watches, 7, 9, 10, 12, 13, 29, 63, 71
and World War II, 12, 13, 39, 40
Ingraham, Dudley, 9, 13, 14, 15, 47, 52, 53, 58, 59, 71, 72
Ingraham, Edward (I), 8, 9
Ingraham, Edward (II), 9, 13, 15, 16, 18, 19, 20, 24, 25, 29, 45, 46, 47, 48, 49, 53, 58, 59, 60
Ingraham, Elias, 7, 8, 14, 15
Ingraham, William S., 9, 14, 24
International Union of Electrical Workers (IUE) Local 260 (Ingraham)
    closed shop, 65
    contracts, 65
    equal pay fight, 65
    executives, 50, 55, 56
    strikes, 65, 67
IUE Local 261 (Sessions)
    closed shop, 67, 68
    membership, 54, 65, 66
    1967 strike, 67, 68, 69
    scapegoating, 67

Japan, 13

Korean War, 13, 60, 63

"Lady Sessions," 58
Landes, David, 3
Laurinburg, NC, 57, 58, 59
Laviero, Kathryn "Kay," 16, 19, 20, 21, 23, 25, 34, 37, 39, 44, 47, 49, 51, 72

Mallory, P. R., 80, fn 9
Manufacturers Association of Connecticut/Hartford County, 15, 53

National Labor Relations Board (NLRB), 43
National Recovery Association
    Code Of Fair Competition for clock and watch industry, 11

Neece, Bret, 59, 60
New Departure Division of General Motors, 8, 43
    United Auto Workers at, 43
New Haven Clock Company, 16

Office of Price Administration, 13

Paternalism, 2, 7, 14, 15, 17, 19, 24, 25, 26, 35, 47, 49, 51, 57
Plainville, CT, 8, 20, 68
Plymouth, CT, 7
Power, George, 19, 22, 28, 50, 53, 56, 65, 71, 72

Reconstruction Finance Corporation (see Sessions Clock Company), 11, 44
Rosenzweig, Roy, 3

Sakowski, Walantyna, 20, 38, 60, 64
Sessions Clock Company
    case shop, 10, 15, 31, 32, 33, 51
    clock assembly, conveyor, 35
    clocks, electric, 9, 11
    clocks, manual, 9
    closing, 65, 67, 71
    division of labor, 1, 2, 17, 23, 27, 31
    during Depression, 7, 11, 12, 15, 19
    early history, 7
    foremen, 51
    layoffs, 15, 19, 23, 36, 52, 68
    MEPCO Division, 57
    management, 2, 4, 14, 66
    mortgage, 57
    motors, 11, 37, 57, 71
    packing, 35, 36
    paternalism, 2, 4, 7, 13, 14, 15, 17, 18, 19, 24, 25, 26, 50
    presswork, 36, 40
    Reconstruction Finance Corporation loan, 11
    skilled labor, 18, 30, 38
    spray room, 38
    subassembly, 33, 35
    taxes owed to Bristol, 11, 57
    timers, 25, 36, 57, 67
    unions, 2, 32, 44, 48, 66, 67
    unskilled labor, 18, 31, 32
    wages, 35, 44, 45, 57, 66, 67
Sessions, John Humphrey, 10
Sessions, William E. (W. E.), 7, 10, 11, 14
Sessions, William Kenneth, Jr. ("Bud"), 16, 18, 25, 26, 44, 45
Sessions, William Kenneth, Sr. (W. K.), 10, 11, 14, 15, 16, 18, 25, 44, 45, 57
Spring, Nelson, 20, 21, 28, 51, 58, 60
Strikes, 2, 17, 50, 52
Switzerland, 12, 16

Taft-Hartley Act of 1947, 53

Tariffs, 12, 13, 52, 60, 68, 71, 72
    See also Hull Reciprocal Trade Agreements
Terryville, CT, 8, 53
Tetro, Robert, 15, 20, 25, 28, 31, 37, 45, 46, 48, 50, 51, 58, 59

United Electrical, Radio, and Machine Workers (UE)
Local 260 (Ingraham), 32
    case shop origins, 45
    communism charges against, 43, 53, 54, 56
    contracts, 48
    executives, 45, 46, 47, 51, 52
    movement shop attitudes toward, 45, 48
    NLRB elections, 45, 54, 55
    portal-to-portal suit, 52, 53
    secession crisis, 54
    strikes, 50, 52
    women's role in, 48, 49, 50
    xenophobia within, 43
UE Local 264 (Sessions), 44
    demise, 43
    executives, 45
    origins, 44
    women's role in, 50, 65, 66
U. S. Time Corporation (Timex), 29, 64
United Electrical Workers, 24, 32, 44

Vietnam War, 60, 64

Watch adjusting, 27, 29, 40
Welch, E. N., 7, 8, 10
Welch Manufacturing Company, E. N., 7, 8
Work cultures, 2, 4, 38, 61
Workers
    attitudes toward unions, 1, 2, 4, 17
    class consciousness, 1, 2, 24, 26, 64
    dating, 21
    employment patterns, 1, 17
    ethnicity, 1, 2, 4, 7, 17, 23, 32, 65
    family strategies, 2, 7, 15, 17, 19, 40, 72
    married men, 15, 17, 23, 24, 27, 28, 37, 40
    married women, 2, 17, 19, 23, 24, 27, 36, 40, 64
    off-the-job activities, 3, 16
    pranks, on-the-job, 37, 38
    race, 1, 2
    single men, 17, 20, 37, 40
    single women, 17, 20, 36, 40
    time consciousness, 3
    wages, 16, 22
    widows, 17
Welfare capitalism. See paternalism. 4, 7, 15, 26
World War I, 24, 35, 43
World War II, 1, 7, 11, 12, 20, 23, 25, 27, 34, 39, 40, 53, 57, 62, 63

Zahavi, Gerald, 17

To Experience Adventures in TIME...
Join the NAWCC

It All Starts with Membership

The National Association of Watch and Clock Collectors, Inc. (NAWCC) is an international nonprofit association serving more than 17,000 members and 150 chapters and dedicated to preserving and stimulating interest in horology, the art and science of time. Our members are enthusiasts, students, educators, casual collectors, businesses, and professionals, who love learning about the clocks and watches they preserve, study, and collect. Members share their interests with other members and establish friendships around the world.

Membership Advantages
- Stay informed with the bimonthly *Watch & Clock Bulletin*, an educational journal, and the *Mart & Highlights*, a buy/sell/news publication.
- Go online for research tools and videos: NAWCC.org features all *Watch & Clock Bulletin* content back to 1943, NAWCC books and instructional videos, and much more for members.

- Buy, sell, and learn at regional buying and selling venues, and attend programs on all aspects of horology.
- Keep in touch with our bimonthly electronic newsletter—*eHappenings*
- Meet terrific people at local and special interest chapters
- Visit for free the National Watch & Clock Museum in Columbia, PA.
- Use your membership for free or discounted admission to over 250 museums and science centers.

Become a member today and begin your exploration of the fascinating world of horology.
Apply online at www.nawcc.org

Become a member today!
Mail this application, apply online at www.nawcc.org, or call 1-877-255-1849 or 1-717-684-8261.

*Required

*Print Name

Company Name (optional)

*Street

*City

*State/Province/Country *Zip/Postal Code

( ) ( )
Ph.: Home Work

( )
Cell

Email

*Former member of NAWCC? ☐ Yes _____ ☐ No
Member Number

/ /
Date of Birth For verification purposes. Required for Youth Membership.

How did you learn about the NAWCC (or from whom)?

Interest: ☐ Clocks ☐ Wristwatches ☐ Pocket Watches ☐ Museum
☐ Other:_____

As an NAWCC member I agree to abide by the standards of fairness and honesty described in the NAWCC Member Code of Ethics, available at nawcc.org.

Send this application with payment to:
NAWCC, Inc., 514 Poplar Street, Columbia, PA 17512-2130
Annual dues:
☐ **Individual $82** (**mailed pubs.) ☐ **Individual $72** (***electronic pubs.)
☐ **Business $150** (**mailed pubs.) ☐ **Associate $20** (***electronic pubs.)
Spouse of member or youth under 18.
☐ **Student $35** (***electronic pubs.)
Please attach copy of student I.D.

Not sure? Try our 4-month membership!
☐ **Introductory $25**
(includes mailed publications)

**mailed publications annual membership includes six issues each of the *Watch & Clock Bulletin* and *Mart & Highlights*.
***electronic publications indicates online versions only.
(Online publications are available to all members.)

Payment:
☐ Check/Money Order (U.S. bank with U.S. address only) ☐ PayPal
☐ Visa ☐ MasterCard ☐ Discover ☐ American Express

Credit Card No. or PayPal Account No.

/
Exp. Date Security code on back of card

Cardholder's Name Amount to be charged

Signature

www.ingramcontent.com/pod-product-compliance
Lightning Source LLC
Chambersburg PA
CBHW060234240426
43671CB00016B/2937